SpringerBriefs in Health Care Management and Economics

Series Editor

Joseph K. Tan, DeGroote School of Business, McMaster University, Burlington, ON, Canada

SpringerBriefs present concise summaries of cutting-edge research and practical applications across a wide spectrum of fields. Featuring compact volumes of 50 to 125 pages, the series covers a range of content from professional to academic. Typical topics might include:

- A timely report of state-of-the art analytical techniques
- A bridge between new research results, as published in journal articles, and a contextual literature review
- A snapshot of a hot or emerging topic
- An in-depth case study or clinical example
- A presentation of core concepts that students must understand in order to make independent contributions

SpringerBriefs in Health Care Management and Economics showcase emerging theory, empirical research, and practical application in health care, health economics, public health, managed care, operations analysis, information systems, and related fields, from a global author community. Briefs are characterized by fast, global electronic dissemination, standard publishing contracts, standardized manuscript preparation and formatting guidelines, and expedited production schedules.

More information about this series at http://www.springer.com/series/10293

Murray V. Calichman

Essential Analytics for Hospital Managers

A Guide to Statistical Problem Solving

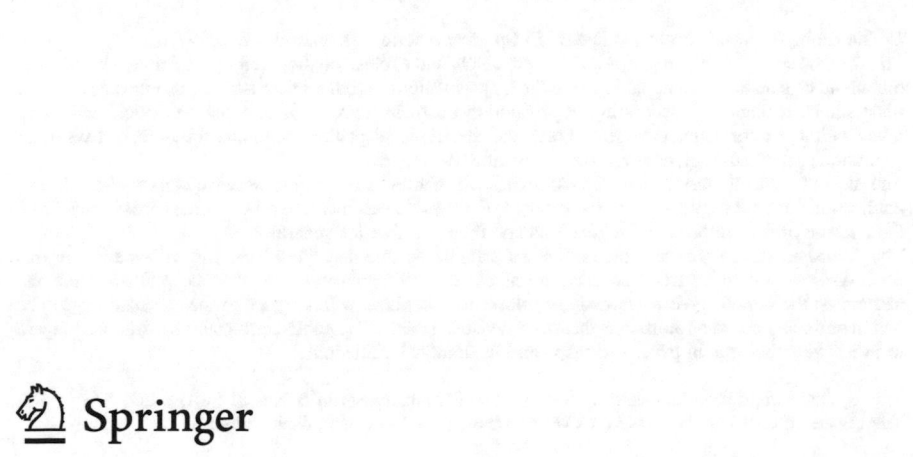 Springer

Murray V. Calichman
Huntington Station, NY, USA

ISSN 2193-1704   ISSN 2193-1712   (electronic)
SpringerBriefs in Health Care Management and Economics
ISBN 978-3-030-16364-8   ISBN 978-3-030-16365-5   (eBook)
https://doi.org/10.1007/978-3-030-16365-5

Mathematics Subject Classification (2010): 62P10, 90C05, 90B99

This Springer imprint is published by the registered company Springer Nature Switzerland AG
The registered company address is: Gewerbestrasse 11, 6330 Cham, Switzerland

Preface

In 1964, I graduated from Rensselaer Polytechnic Institute (RPI) with a Bachelor's Degree in Management, the first year that a bachelor's degree was conferred in that discipline at the school. In my junior year, I was allowed to take an evening graduate course in Operations Research and, academically, everything changed for me. I soon transferred from the school of mathematics to the school of management.

I recently retired after providing management services to hospitals for almost 50 years; 2 years as a management engineer with the Nassau-Suffolk Hospital Council on Long Island, 35 years as President of GBC Consulting Corporation and the last 11 years as Director of Operational Research & Analytics at St. Francis Hospital/Catholic Health Services of Long Island.

I saw my role as a management scientist/engineer in the hospitals where I was privileged to work as (a) one of support—supporting the efforts of front-line personnel, as well as hospital management, by helping to create a work environment where "success" could be achieved and (b) as an agent for change.

It is for these purposes that I have written this book—as a continuing attempt to improve hospital operations. I selected 12 topics that I thought would be of most interest to the reader and sequenced them in order by mathematical/statistical complexity. The book reviews simple and more advanced methodologies in a "How to" format by introducing actual hospital problems and the methodologies used to solve them.

The book provides practical applications of statistical and mathematical concepts. Each chapter discusses a key component of hospital operations, be it having enough linen on hand on the nursing units, having sufficient medical surgical supplies when and where they are needed, minimizing the cost of holding and ordering those supplies, determining the appropriate number of beds required given various circumstances, devising and implementing productivity and cost reporting systems, developing consistent staffing criteria from different sources, forecasting facility requirements from historical data, developing optimum pricing models to maximize hospital profitability, and last, devising elective patient schedules to fully utilize hospital beds and eliminate patient overcrowding.

The last two chapters provide innovative use of linear programming; introducing the means to maximize hospital profitability through pricing changes and the means to eliminate overcrowding in hospitals by balancing the demand for beds each day of the week; one where the distribution of lengths of stay by surgical category becomes the prime component in the determination of the optimum patient schedule.

The book is structured by degree of mathematical complexity. The very first chapter involves no mathematics at all, but provides a foundation for improving the efficiency of the work environment—Work Simplification. The chapters following proceed to offer problem-solving methodologies, utilizing basic statistical concepts, such as "means", "standard deviations", etc., more advanced statistics, such as "Poisson distributions", and concludes with chapters utilizing computer simulation and linear programming.

Although this book will not allow managers to master the topics presented, it will provide an overall awareness of how statistics/mathematics can be utilized to solve hospital problems.

It is my sincerest hope that the reader will experience some of the pleasure I felt working in an industry where problem-solving not only benefited the hospitals where I worked, but their patients as well. Working in healthcare provided a very rewarding career.

Huntington Station, USA   Murray V. Calichman

Acknowledgements

In one way or another, everything written in this text had its beginning in one of three books; one of which I have had with me since my undergraduate years at Rensselaer Polytechnic Institute, the second since taking an advanced statistics course at the Bernard Baruch School of Business at City College and the third a book on simulation I was fortunate to obtain many years ago.

The first book is entitled, "Operations Research—methods and problems" written by Maurice Sasien, Arthur Yaspan and Lawrence Friedman, copyright 1959 by John Wiley & Sons, Inc. It was the textbook for both my introductory course in Operations Research at RPI and the Operations Research course in graduate school.

The second (and foundation) book is entitled, "Statistical Analysis—Second Edition", written by Samuel B. Richmond, copyright 1957 and in 1964 by The Ronald Press Company. Simply stated, I refer to this book's chapters and accompanying tables all the time.

The third book is entitled, "Introduction to Simulation Modeling Using GPSS/PC" written by James A. Chisman, copyright 1992 by Prentice-Hall, Inc. Although I had taken a course given by IBM on GPSS in the early 1970s, and began to use simulation soon thereafter, this book was a most valuable addition in that it begins with a "refresher" on statistical analysis and then details the various commands required by this simulation language, along with practical problems and solutions.

I am indebted to and extremely grateful for each of these authors.

I would like to acknowledge and thank Dr. Akram Boutros, the CAO at SFH, who hired me as the Director of the Department of Operational Research and Analytics (DORA) in 2007, for creating this multi-disciplinary department, one of the first of its kind in Healthcare. I also thank Ms. Ruth Hennessey, who succeeded Dr. Boutros as CAO, for maintaining the department's viability for the next 10 years. How fortunate I was to work for two such outstanding executives.

I also want to acknowledge the direct and indirect contributions made to this book by the members of the DORA department at St. Francis Hospital, each a seasoned professional in his/her field.

Last, my thanks to Thomas B. Doolan, who was the assistant administrator at St. Charles Hospital when I first went to work for the Nassau-Suffolk Hospital Council in 1970, and who provided me with consulting assignments continuously for the next 30 years, as administrator of two hospitals in New Jersey and as the system administrator for a number of hospitals on Long Island.

Contents

Chapter 1
Work Simplification—A Method to Improve Processes

A Work Flow Diagram, A Little Logic

Work Simplification is the systematic use of common sense in the quest for better and easier methods of accomplishing the work.

As a tool to combat the increasing costs associated with the performance of each activity at the job site, every employee should utilize the Work Simplification approach to assist in analyzing his/her job assignment. As a body of knowledge, work simplification contains principles useful in the discovery of better ways of carrying out assigned tasks. It is based upon the proposition that there is "one best way" of performing work. Even though such an idealistic goal may never be realized, attempts to strive toward this one best way will result in improvements that save time and money.

Work Simplification means making improvements by:

1. **Eliminating** unnecessary jobs or parts of those jobs.
2. **Combining** parts of the job.
3. **Re-arranging** the sequence of parts of the job.
4. **Simplifying** the necessary parts of the job.

The pattern for achieving these results is first to select the job to be improved. The jobs to select initially are those that take the most time, those that bottleneck other operations and those that are the most costly to perform.

The second step in the process is to break down the job in detail, most usually by using a simple Flow Chart, see Fig. 1.1, and to question each component of the job. Most jobs consist of some, or all, of five components.

1. **Operations** (the actual doing, usually represented by a large "O" in more sophisticated flow charting)
2. **Transportations** (the movement of material from one station to another, usually represented by a large arrow)
3. **Inspections** (the verification that the operations and transportations were done correctly, usually represented by a large square)

© The Author(s), under exclusive license to Springer Nature Switzerland AG 2019   1
M. V. Calichman, *Essential Analytics for Hospital Managers*,
SpringerBriefs in Health Care Management and Economics,
https://doi.org/10.1007/978-3-030-16365-5_1

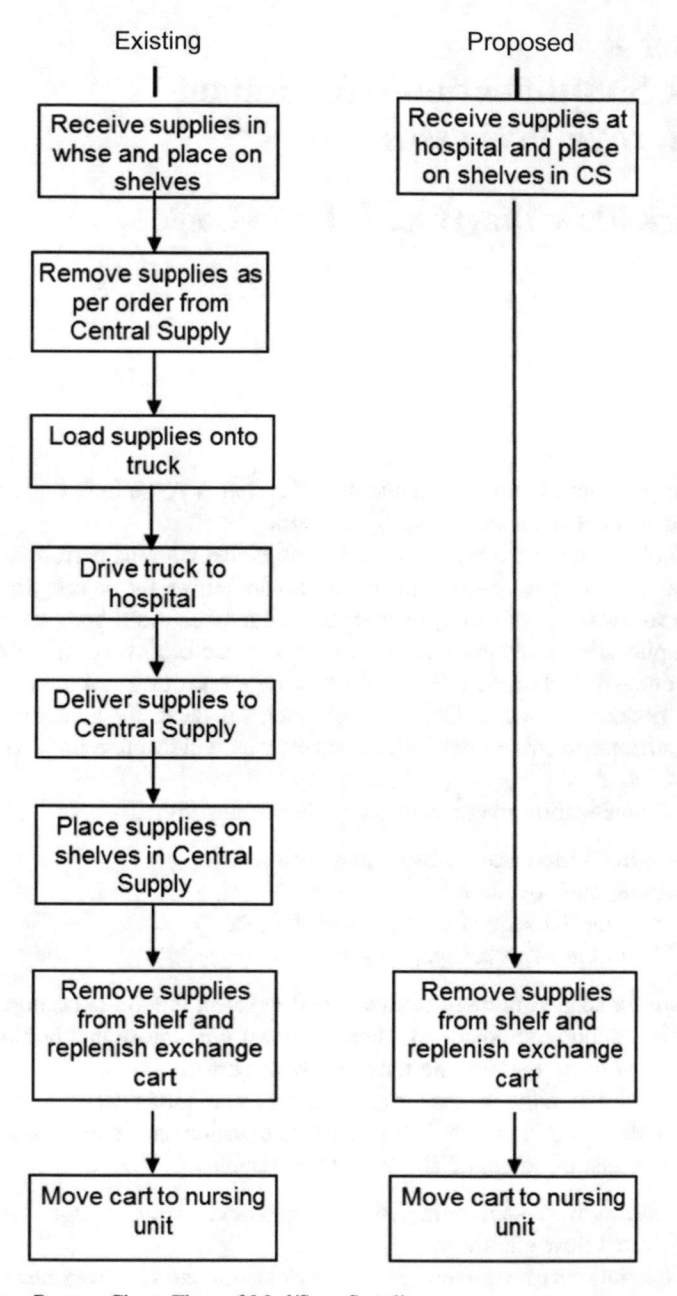

Fig. 1.1 Flow Process Chart: Flow of Med/Surg Supplies

4. **Delays** (the wait for something required to occur, usually represented by a large "D")
5. **Storages** (the temporary or final location of material, usually represented by a large triangle).

The object of this process is to minimize the time and cost to perform the job by reducing the job's required number of steps. Thus, it is imperative to challenge each job with the question, *Why? Why* is the job done at all? *What* are the ramifications of not doing the job? *Why* is each operation of the job done, and finally, *why* is each transport, inspection, delay and storage activity performed?

To aid in questioning the "Why" of the job, and each detail of the job, use each of the five prompters that follow:

1.	What is done?	Why is it done at all? What else could be done to accomplish the same results?
2.	Where is it done?	Why is it done there? Where else could it be done?
3.	When is it done?	Why is it done then? At what other time could it be done?
4.	Who does it?	Why does this person do it? Who else could do it?
5.	How is it done?	Why is it done this way? In what other way could it be done?

After the questioning process is completed, and appropriate answers agreed upon, a new method for doing the job will evolve. Figure 1.1—Flow of Med/Surg Supplies—is a sample comparison of what a new process may look like versus the existing process when a Work Simplification project is completed. You will note that the number of steps in the process was reduced from eight to three.

The new method should be presented to the individual or group of individuals empowered to approve changes in job method. If approved, the new method should be implemented. If not approved, the reasons why not approved should be explored and the entire process repeated.

The application of Work Simplification will empower employees to make improvements that directly impact their work and result in cost savings for the institution. Also, it will enable the program participants to gain greater job satisfaction in the knowledge that each is doing his or her part to contain costs and make the institution even more financially viable.

Chapter 2
The Linen Multiplier

Derivation of the Linen Multiplier—Simple Mathematics, Simple Logic

The Linen Multiplier is the number of days of linen a hospital should have in circulation. It is "x" times the amount of each linen item used each day, on average. This chapter illustrates how to derive "x" for your hospital.

Hospital linen plays a major role in the comfort of hospital patients. Shortages are to be discouraged. Unfortunately, many hospitals suffer unnecessary and prolonged periods of less than the optimum quantity of linen in circulation.

The life of any linen item is dependent upon many factors; amongst them are the quality of the linen (threads per inch), the chemicals in and the temperature of the water during laundering, pilferage, number of uses, environment of the area where stored, how used, etc. Although it is incumbent upon the hospital to control all of these factors in order to extend the life of each linen item, item life has much to do with ultimate linen costs but little to do with the quantity of each item required each day. With appropriate usage records, the number of each linen item required each day will become known over time.

A 24-h linen cart(s) should be exchanged each day in those areas of peak linen usage, such as the nursing units. A known quantity of each linen item should be placed on each cart. Calls for additional linen should be recorded during the day, as should the number of items returned to the linen room each day on the depleted carts. Assume, for example, that 100 large sheets were delivered on the linen cart to nursing unit 2A, that there are calls during the day for an additional two dozen large sheets and the depleted cart was returned to the linen room the following day with a dozen large sheets. Linen usage for that day, for that unit, would be determined at 112 large sheets (100 + 24 − 12). In a similar manner, the linen usage for all areas can be determined. In lieu of recommended daily usage records, the linen manager could conduct a two-week data collection period some time during the hospital's busy months in order to determine the hospital's daily linen needs.

Before proceeding any further, it should be pointed out that there is nothing sacrosanct about exchanging linen carts first thing each morning. One hospital practically eliminated the call for additional linen during the 24-h day by moving the time to exchange linen carts from approximately 7:00 a.m. each day to

M. V. Calichman, *Essential Analytics for Hospital Managers*,
SpringerBriefs in Health Care Management and Economics,
https://doi.org/10.1007/978-3-030-16365-5_2

approximately 7:00 p.m. Data indicated that most calls for linen were made before the evening shift left for the day. The evening nurses were "stocking up" so the night nurses would not be left without sufficient linen. Data also indicated that most, if not all, of the called for linen was returned the next morning. With the cart exchange moved to the early evening hours, the evening nurses had no need to "stock" linen (or overstock, as was really the case) for the next shift, as the linen carts were stocked with a 24-h supply. And, if the supply of linen on the unit were to run low, it would occur during the day shift. Thus, there was no need for additional stocking; for on the day shift extra linen was available in a matter of minutes.

Once the quantity of each linen item required each day is calculated, it is easy to determine the amount of linen required to be in circulation. In today's environment, most hospitals utilize an outside laundry to pick-up soiled linen, to perform the laundering and to return the clean linen to the hospital. Because these outside laundries pay their employees minimum wages, it has become increasingly more difficult for hospitals to cost justify laundry operations on-site. In any event, the most frequently used laundry model today is a hospital serviced by an outside laundry.

Let us assume that the outside laundry has a six-day per week schedule. Every morning, Monday through Saturday, it delivers the laundry picked up the previous day (and laundered that day) and picks up the soiled linen.

To determine the total linen required in circulation for this particular model, let us review the phases of the hospital linen cycle. The cycle consists of the following steps:

Day 1. In use on beds, or elsewhere
Day 2. Removed from beds, or soiled elsewhere, and sent to dirty linen area
Day 3. Picked up by outside laundry service (and laundered)
Day 4. Returned by outside laundry service

Setting up a matrix with the days of the week across the top and the cycle steps on the left will lead us to the required linen multiplier for this model, as such (where each letter stands for 1 day of required linen) (Table 2.1).

Table 2.1 The linen process by day of week

Step	M	T	W	TH	F	SA	SU	M	T	W	TH	F	SA	SU	M	T	W
In use	A	B	C	D	A	B	C	E	D	A	B	C	E	D	A	B	C
Soiled		A	B	C	D	A	A/B	C	E	D	A	B	C	C/E	D	A	B
Picked-up			A	B	C	D		A/B	C	E	D	A	B		C/E	D	A
Returned				A	B	C		D	A/B	C	E	D	A		B	C/E	D
Storage									B	C	E	D	A				E

As is seen in the above table, the first day's supply of linen (A) is used on Monday, soiled (and in the dirty linen area) on Tuesday, picked up by the outside laundry on Wednesday and returned to the hospital on Thursday (to be available for use on Friday). Note that four days of linen is required to satisfy the linen demand through the first Thursday, being that the first day's supply (A) has only returned to the hospital on that day. These four days of linen supply rotate through each component of the linen cycle through Saturday.

On Sunday, however, the laundry neither picks up, launders nor delivers clean linen. Thus, as the above indicates, on the first Sunday daily supplies A and B are sitting in the hospital's dirty linen area, daily supply C is in use in the hospital and D is sitting in the laundry. On Monday, therefore, it is necessary for the hospital to have available a fifth day supply of linen, or E.

It is also seen from the matrix that there are days when the hospital will have an extra day's supply of linen. This has to be a consideration when planning storage space. For the model discussed, the linen room would have to provide space for one day of linen storage plus the carts that will deliver the linen in use that day. As is also apparent, the table reaches a steady state with five times the daily usage.

Keep in mind, however, that five days of linen supply (A–E) should be viewed as the *minimum* quantity of linen required. Due to spikes in usage, potential problems with any of the linen phases, etc., there should be a sixth day of linen, increasing the linen multiplier for this hospital from five to six and increasing required storage space by another day's supply. For a hospital that uses 1200 large sheets per day, for example, there should be 7200 large sheets (600 dozen) *in circulation*.

Also, the hospital should divide its required linen purchases by 12 and place a standing order for the various linen items with its linen vendor for 1/12th of its replenishment needs each month (This is where the life cycle of each item comes into play). When received, the linen should be inspected, stamped, laundered and placed immediately in circulation (thereby providing between five and six days supply of linen in circulation at all times). There is no advantage to be gained by placing the linen in storage. If the *life* of a large sheet in the above example is 3 years, then 1/3 of 600 dozen, or 200 dozen large sheets should be replenished each year. Thus, a standing order should be placed for delivery of 17 dozen large sheets, amongst other items, each month (200/12).

The linen multiplier is thus dependent upon two factors; the steps in the linen cycle and the (outside) laundry schedule. An increase in the number of steps or a decrease in the workweek will result in the need for a greater multiplier. Conversely, a decrease in the number of steps or an increase in the workweek will result in the need for a lesser multiplier.

Chapter 3
The Emergency Room Holding Area

Averages, Standard Deviations and Confidence Levels

For a number of years now, hospitals have been plagued by inpatients being held in the emergency room awaiting inpatient beds (Refer to Chap. 12- Use of Linear Programming to Eliminate Hospital Overcrowding in Hospitals, as a means of eliminating the overcrowding problem. This chapter is written for those hospitals that function with overcrowding but need to right size their Emergency Room holding area.).

In these hospitals there are no beds available to accommodate these patients on the floor-nursing units. Years ago these patients were placed in "hall" beds on the nursing units. For a number of reasons, this practice was discontinued and patients are now routinely held overnight (perhaps for many nights) in an emergency room holding area.

This transition has caused many hospitals to re-allocate their emergency room space. Many hospitals carved out space to provide an "inpatient" unit within the confines of the emergency room area. Once it is decided to provide this space, the question then becomes, "How many beds are required to accommodate these patients?"

This chapter will attempt to answer that question and to provide some insight into the use of basic statistics.

I received an urgent call from a hospital's Director of Nursing one day many years ago. The architect that the hospital had hired to "re-allocate" emergency room space was due in one hour. Her "committee" had concluded that the hospital would require space for four (4) beds to accommodate inpatients in the yet-to-be-allocated space for the emergency room holding area. The Director had a "gut" feeling that this conclusion was incorrect but did not have the time (or perhaps the knowledge) to quickly come up with a better answer for her meeting.

The only data that was available to her was the number of inpatients that were held overnight (the overnight census) in the emergency room from a recent 100-day period (see Fig. 3.1). A review of the data revealed that the "average" census of admitted patients still in the emergency room at midnight was four. Evidently, that was the reason for the committee to conclude that four beds were required for the

© The Author(s), under exclusive license to Springer Nature Switzerland AG 2019   9
M. V. Calichman, *Essential Analytics for Hospital Managers*,
SpringerBriefs in Health Care Management and Economics,
https://doi.org/10.1007/978-3-030-16365-5_3

EMERGENCY ROOM - INPATIENT CENSUS

DAY	NUMBER	DAY	NUMBER	DAY	NUMBER	DAY	NUMBER
1	0	26	3	51	3	76	1
2	0	27	2	52	5	77	1
3	3	28	4	53	6	78	6
4	2	29	9	54	6	79	6
5	0	30	10	55	6	80	5
6	5	31	9	56	6	81	6
7	4	32	9	57	2	82	5
8	4	33	4	58	1	83	6
9	1	34	1	59	1	84	4
10	1	35	1	60	1	85	5
11	1	36	1	61	2	86	5
12	1	37	3	62	2	87	4
13	2	38	5	63	2	88	1
14	0	39	4	64	0	89	1
15	0	40	4	65	0	90	8
16	0	41	4	66	5	91	8
17	3	42	3	67	7	92	7
18	5	43	2	68	7	93	4
19	7	44	7	69	7	94	5
20	8	45	7	70	9	95	3
21	8	46	7	71	9	96	0
22	11	47	9	72	5	97	0
23	1	48	11	73	5	98	0
24	1	49	6	74	4	99	3
25	0	50	4	75	2	100	6

Total	400
Average	4.00
Standard Deviation	2.93
95% Confidence Extreme	
(Average + 1.64 Stnd Dev)	8.81

Fig. 3.1 Emergency room inpatient census at midnight

holding area. However, unlike situations where the "average" may dictate decision making, in cases like this the average does not suffice.

As the census statistics indicate, there were nights when the Emergency Room had zero inpatients waiting for available floor beds and nights when there were up to ten and eleven patients waiting for those beds.

What should be done is to size an area to accommodate all of those patients 95% of the time. Stated somewhat differently, the area should be able to accommodate those patients awaiting floor beds 95 out of every 100 nights. As such, the average of four beds would not suffice for the many nights when 5, 6, 7 or more patients are held. Therefore, in addition to the "average", another statistic of equal importance must be derived; the "standard deviation".

Although not correct, it is often said that the standard deviation is the average difference between each occurrence and the group average (Actually it is the square root of the average differences between each occurrence and the sample mean squared). And, as any good statistical text book will indicate, the average plus and minus 1.0 standard deviation will provide a range of numbers that contain a little more than 68% of all occurrences; the average plus and minus 1.5 standard deviations will provide a range of numbers that contain almost 87% of all occurrences, the average plus and minus 2.0 standard deviations will provide a range of numbers that contain almost 95.5% of all occurrences, etc.

The numbers of standard deviations most usually used in this type of problem are 1.64 and 1.96, providing 90 and 95% respectively of all occurrences. Refer to the area under the normal distribution shown in Fig. 3.2 to get a better idea as to what this means. Thus, if one wanted to provide holding space for the number of beds that would satisfy 95% of the nights inpatients are held in the Emergency Room, one would add and subtract 1.96 standard deviations from the average.

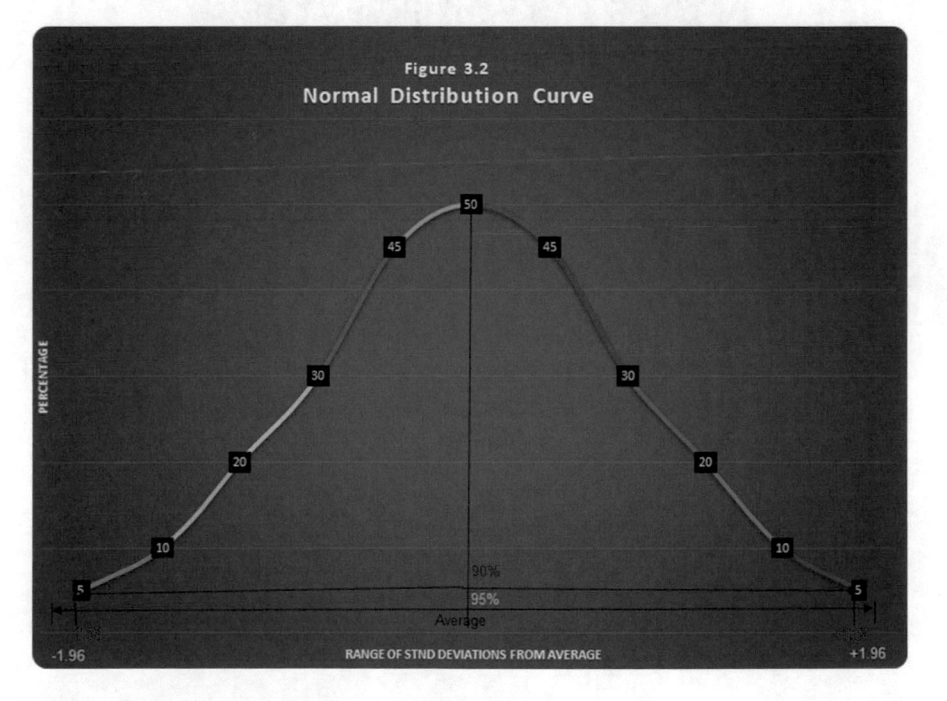

Fig. 3.2 Normal distribution curve

DISTRIBUTION TABLE

INPATIENTS WAITING IN BEDS	OCCURRENCES	%	CUMULATIVE %
0	12	12.0%	12.0%
1	16	16.0%	28.0%
2	9	9.0%	37.0%
3	8	8.0%	45.0%
4	12	12.0%	57.0%
5	12	12.0%	69.0%
6	10	10.0%	79.0%
7	8	8.0%	87.0%
8	4	4.0%	91.0%
9	6	6.0%	97.0%
10	1	1.0%	98.0%
11	2	2.0%	100.0%
	100		

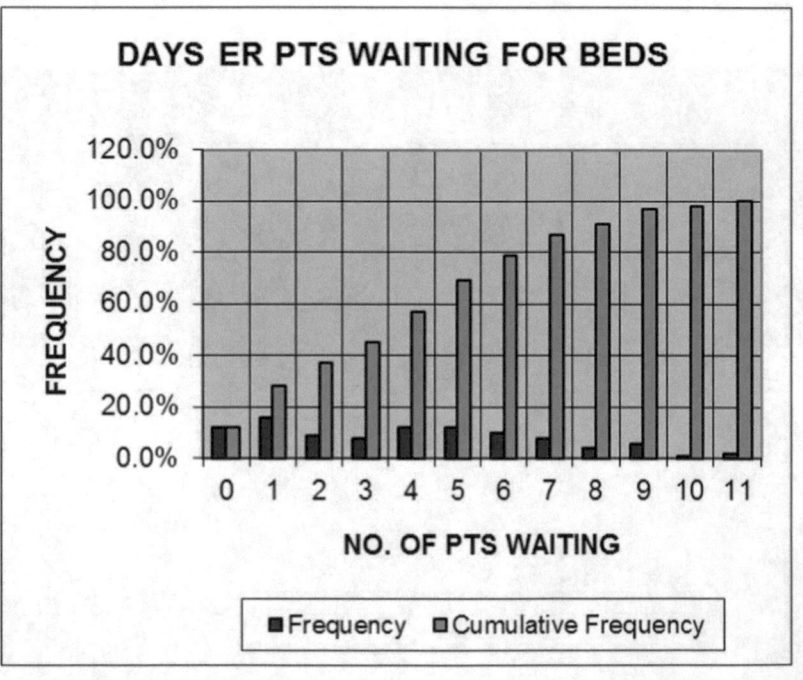

Fig. 3.3 Histogram data and graph

That would mean that 2.5% of all occurrences would fall below the range covered by the average −1.96 standard deviations, and 2.5% of all occurrences would fall above the range covered by the average +1.96 standard deviations. However, since having beds for too few patients is of no concern, the number of standard deviations used in the calculation can be reduced to 1.64 as the only concern is with not having a sufficient number of beds for 5% of the time (again, there is no concern when there are but a few patients overnight in the ER).

In the example above, it was concluded that the architect should find space for 9 beds (the average of 4.0 plus 1.64 times the standard deviation of 2.93, or 8.8 beds). Or, stated another way, with 9 beds there is 95% confidence that each night there will be sufficient beds to accommodate all inpatients waiting in the Emergency Room for floor beds.

With the advent of spreadsheet applications, the mathematics needed to perform these calculations is quite easy. One simply has to stipulate a range of numbers (occurrences) and request the appropriate statistic. I simply asked for three successive statistics from an Excel spreadsheet; total (sum), average (average), and standard deviation (Stnd Dev) and then summed the average with the product of 1.64 times the standard deviation.

Another option available to determine the appropriate number of beds is to align the data in a Histogram, or cumulative frequency distribution, as shown in Fig. 3.3. Note that 9 beds to accommodate nine or less patients each night would be sufficient 97% of the time; consistent with the results achieved by using only the "average" and "standard deviation" statistics.

Chapter 4
Determining "Par" Levels for Patient Supply Items

Averages, Standard Deviations and Confidence Levels

There are a number of different distribution systems used by hospitals in moving supplies from the General Stores/Central Sterile areas to the point of use areas. Many hospitals use either one or a combination of these three supply distribution systems:

1. Traditional Par Levels
2. Exchange Cart Systems
3. Double Bin Systems

Most *traditional par level system*s are characterized by an on unit (supply cabinet) stock of supply items. Usually, individuals from the distribution center (General Stores/Central Sterile) travel to the various nursing units at the same time each day, clipboard in hand, and inventory each unit's supply cabinet. All items required to restore stock levels to par (a 24 h supply) are noted. The individuals return to the distribution center, "pick" the items required to restore the units to their customized par levels, return to the nursing units later in the day and place the new items in their appropriate locations in the supply cabinets. When this service is *not* provided each day, problems usually arise inasmuch as there simply is not sufficient space in the supply cabinet to store an additional day's supplies (a six-day/week operation), never mind an additional two days' supplies (a five-day/week operation). Alternate space is sometimes found but, oftentimes, not communicated to the weekend nurses. It should be mentioned that many hospitals have upgraded to "intelligent" or "automated" cabinets, especially for the more expensive supplies, including pharmaceuticals, which alert the source department when refills are necessary. Although there are labor savings associated with these cabinets, the motivation is usually to provide better control and security.

The *exchange cart system* evolved from the par level system and is similar in many respects. The major difference is that the nursing unit's supply cabinet is on wheels—the exchange cart. Also, a "sister" cart exists for each nursing unit in the distribution center. The "sister" cart is replenished with supplies in the distribution center (each item being brought up to its par level), stored in the distribution center overnight and delivered to the nursing unit the following day. The depleted cart is returned to the distribution center with the process repeating itself each day.

The *double bin system* is the least labor intensive of the three distribution systems discussed and is the newest. However, it also requires the most stock on the nursing units. This system was brought to the United States a number of years ago from Europe. In the double bin system the supplies are usually stored in customized baskets within racks on the nursing units. Each item's stock is divided in two separate sections of a basket/drawer, the front and rear sections.

The individual from the distribution center travels to each nursing unit, each day, Monday–Friday, and opens each basket/drawer. On days when the front section is devoid of supplies, the worker scans the section's bar code (mounted on the vertical divider in each section) into a portable reader. When all empty sections of the baskets are scanned, the worker returns to the distribution center, downloads the scanned information to the computer and prints the supply items required (in geographically-stored sequence). There, workers "pick" the supply items required (a small fraction of the items required to support either the par level system or the exchange cart system, with resulting savings in labor) and deliver them to the appropriate locations on each unit's racks and baskets.

What all three supply distribution systems have in common is the need to determine the quantity of the different supply items required for each nursing unit. The mathematics in making this determination are quite similar to the mathematics required in Chap. 3; the use of averages and standard deviations. Again, a review of data is of the utmost importance. Daily usage records are the source documents for the calculations. Refer to the usage statistics for medium stockings as it appears in Fig. 4.1.

Once again, the average simply does not suffice. If the average number of each supply used were to be placed in the nursing unit's supply closet, exchange cart or cabinet, it is obvious from a review of the usage statistics that many calls would be placed by the nursing units to the distribution center each day, necessitating much additional labor, delays in obtaining supplies, etc. Check to see how many days the average of four medium stockings simply would not suffice. There are 43 days in the sample of 100 days when the usage would be greater than four stockings and a greater number of days when anxious nurses might call down for additional stockings because all four had been used.

For the usage statistics given, nine medium stockings would be the desired number if management were to decide that it wished to maintain sufficient stock to satisfy usage demand 95% of the time (4.0 [Average] + 1.64 × 2.93 [Stnd Dev]). For the double bin system, which requires an even numbered quantity, a total of ten or even twelve items (depending upon bin space) would suffice; which would be divided respectively into sections of 5 or 6. It is suggested that the space allocated for each item be fully utilized. Thus, since bin space is dictated by the length and width of the item, it will be advantageous to add quantity if the height of the item permits; bringing the quantity to at least twice the designated amount but no more than four times the designated amount.

DAY	NUMBER	DAY	NUMBER	DAY	NUMBER	DAY	NUMBER
1	0	26	3	51	3	76	1
2	0	27	2	52	5	77	1
3	3	28	4	53	6	78	6
4	2	29	9	54	6	79	6
5	0	30	10	55	6	80	5
6	5	31	9	56	6	81	6
7	4	32	9	57	2	82	5
8	4	33	4	58	1	83	6
9	1	34	1	59	1	84	4
10	1	35	1	60	1	85	5
11	1	36	1	61	2	86	5
12	1	37	3	62	2	87	4
13	2	38	5	63	2	88	1
14	0	39	4	64	0	89	1
15	0	40	4	65	0	90	8
16	0	41	4	66	5	91	8
17	3	42	3	67	7	92	7
18	5	43	2	68	7	93	4
19	7	44	7	69	7	94	5
20	8	45	7	70	9	95	3
21	8	46	7	71	9	96	0
22	11	47	9	72	5	97	0
23	1	48	11	73	5	98	0
24	1	49	6	74	4	99	3
25	0	50	4	75	2	100	6
Total			400				
Average			4.00				
Standard Deviation			2.93				
95% Confidence Extreme							
(Average + 1.64 Stnd Dev)			8.81				

Fig. 4.1 Use of medium stockings

Also note, that if circumstances (usually staffing) mandated a Monday, Wednesday, Friday delivery schedule, then considerably more than a one-day supply (par) would be required. Inasmuch as the M, W, F schedule delivers the necessary supply items every other day with the exception of Friday to Monday, a three day span, this schedule requires the need for three times the daily par at a minimum, or for medium stockings, 3×9 or 27 (28) stockings as a minimum and 4×9 or 36 as a maximum.

Chapter 5
Merging Staffing Standards with a Patient Classification System

Weighted Averages

For years, many hospitals maintained two separate and disparate systems for determining staffing requirements; through budgeting, starting with patient to nurse ratios and forecasted patient volumes, and through a patient classification system.

Although there are a number of ways in which required staffing can be accomplished, allowing nurse management to set the parameters usually works best. Determine the staffing required at full census by converting the unit's staffing needs (fte's) to the number of worked hours required each day. Then divide the worked hours required each day into two categories; a fixed component and a variable component. The fixed component is usually comprised of the head nurse and unit secretaries; the variable component for all other nursing personnel.

Then divide the variable hours required each day by the unit's census at peak occupancy to derive the number of variable nursing hours required per patient day. To this add the fixed component for a two-week period (the payroll period) to derive a so-called production standard that will be used to measure the unit's utilization of labor. These numbers also provide the *basis* for the department's budget.

Let us assume, for example, the following staff requirements for a 32 bed medical-surgical nursing unit each day of the week:

	Fixed	Variable
Days	1.0 head nurse	4.0 registered nurses
	1.0 unit secretary	3.0 licensed practical nurses
		2.0 aides
Evenings	1.0 unit secretary	2.0 registered nurses
		2.0 licensed practical nurses

(continued)

© The Author(s), under exclusive license to Springer Nature Switzerland AG 2019

M. V. Calichman, *Essential Analytics for Hospital Managers*,
SpringerBriefs in Health Care Management and Economics,
https://doi.org/10.1007/978-3-030-16365-5_5

(continued)

		1.0 aide
Nights		2.0 registered nurses
		2.0 licensed practical nurses
		1.0 aide
Total Fte's	3.0	19.0

Staff h/day 22.5	142.5 (19 × 7.5)
Variable h/Pt. day	4.45 (142.5/32)
Fixed h/PP	315.0 (3 × 7.5 × 14)

Thus, the standard for this nursing unit would be established at 315 fixed hours per pay period plus a variable of 4.45 h per patient day. Now, if the nursing unit experiences 90% of occupancy over a two week pay period, or 403 patient days, the unit would earn 315 fixed hours plus 1793 variable hours (4.45 × 403), or 2108 h in total. Any hours worked over that amount would be judged excessive, or overstaffed hours; any hours worked under that amount would be judged as understaffed hours.

The monitoring of productivity performance, via the use of department work standards, should be an operational tool utilized in every hospital (see Chap. 6— Productivity and Cost Application). It is a means of keeping actual expenses in line with actual revenue. Comparing actual expenses with budgeted expenses only, has proved detrimental to far too many hospitals; hospitals that have experienced significant decreases in their expected revenue.

As mentioned at the beginning of this chapter, some hospitals also maintain a patient classification system. Unfortunately, in most hospitals, there is a long-term problem in aligning the two programs. Over time, both programs should provide comparable results. In order for this to occur, however, the patient classification program must be adjusted to comply with overall staffing standards. In other words, the hours of nursing care per *average* medical-surgical patient in the patient classification program must be adjusted to equal the overall nursing hours required per patient as per the department standard (upon which the budget is predicated).

Let us, therefore, go back to the above mathematics for a moment and convert the fixed nursing component to variable hours per patient day. The 22.5 h of fixed nursing care required per day divided by the maximum of 32 patients per day yields another 0.70 h per patient day, or an overall combined total of 5.15 nursing care hours per patient day.

Patient classification programs are usually divided into four or five classes; ranging from the easiest to care for patient to the most difficult to care for patient. Each classification has a description of the type of patient falling within it. The values given to each classification may look something like the following (Table 5.1).

The nurses are to use these various values to determine the care required each day on their respective nursing floors. They are to multiply the value in each class

Table 5.1 Nursing hours per patient day by class

Class I	2.0 nursing care hours per patient day
Class II	4.0 nursing care hours per patient day
Class III	6.0 nursing care hours per patient day
Class IV	8.0 nursing care hours per patient day
Class V	24.0 nursing care hours per patient day

by the number of patients in each class and then divide the combined total value by 7.5 to determine the number of nursing personnel required for the day. However, determining staffing by this means will, most assuredly, be in conflict with the budget. Although the patient classification system is useful in providing the rationale to adjust staffing internally from day-to-day, over time, it simply must conform to the budgeted staffing plan.

In order to achieve this conformity, the absolute numbers in the patient classification system must change. Note that the class II patient requires twice the nursing care of a class I patient (4.0 h vs. 2.0 h), a class III patient requires three times the care of a class I patient, a class IV patient four times the care and a class V patient 12 times the care of a class I patient.

These relationships can be used to make certain that the average value of the patient classification program is adjusted to equal 5.15 h per patient day as established in the staffing budget.

As mentioned many times in the previous chapters, historical data has a primary role in the necessary calculations. A year's review of the classification system might yield the following data (Table 5.2).

There is now sufficient data available to determine the new values for the patient classification system.

Assuming "x" is equal to the new value of class I patients; then 20% of all patients (representing class I patients) times the new value for the class I patients plus 25% (representing class II patients) times two times "x" (the relationship between class II patients and class I patients) + 40% (representing class III patients) times three times "x" (the relationship between class III patients and class I patients) + 12% (representing class IV patients) times four times "x" (the relationship between class IV patients and class I patients) + 3% (representing class V patients) times twelve times "x" (the relationship between class V patients and class I patients) must equal 5.15 (the budgeted staffing ratio), or

$$0.20\,(x) + 0.25\,(2x) + 0.40\,(3x) + 0.12\,(4x) + 0.03\,(12x) = 5.15 \qquad (5.1)$$

Table 5.2 Percent of patient days by class

Class I	20% of patient days classified as I
Class II	25% of patient days classified as II
Class III	40% of patient days classified as III
Class IV	12% of patient days classified as IV
Class V	03% of patient days classified as V

Table 5.3 Adjusted nursing hours per patient day by class

Class I	1.88 nursing care hours per patient day (x)
Class II	3.76 nursing care hours per patient day (2x)
Class III	5.64 nursing care hours per patient day (3x)
Class IV	7.52 nursing care hours per patient day (4x)
Class V	22.56 nursing care hours per patient day (12x)

Table 5.4 Nursing care hours per patient day with a % shift by class

Class I	$1.88 \times 20\% = 0.376$
Class II	$3.76 \times 15\% = 0.564$
Class III	$5.64 \times 50\% = 2.820$
Class IV	$7.52 \times 12\% = 0.902$
Class V	$24.00 \times 03\% = 0.720$
New average	$=5.382$

$$\text{Or} \quad 0.20\,(x) + 0.50\,(x) + 1.20\,(x) + 0.48\,(x) + 0.36\,(x) = 5.15$$
$$\text{Or} \qquad\qquad\qquad\qquad\qquad\qquad\qquad\qquad\qquad 2.74\,(x) = 5.15$$
$$\text{Or} \qquad\qquad\qquad\qquad\qquad\qquad\qquad\qquad\qquad\qquad (x) = 1.88$$

As the above indicates, "x" equals 1.88. Thus, the patient classification program should be adjusted to the following values (Table 5.3).

Over time, this adjustment will greatly assist the hospital in having both programs provide management with similar results as to the number of staff required.

And, once this baseline is established, the future distribution of patients into the pre-set five classifications will enable the hospital to determine the *increase* in nursing staff required as the average medical-surgical patient becomes more acutely ill. For example, a shift of 10% of patients out of class II and into class III would/should result in increasing the overall department work standard from 5.15 to 5.38, as the following calculations indicate (Table 5.4).

Whereas it is initially important that the patient classification values conform to the staffing budget, such that the nursing care hours required per patient day is equivalent in both systems, in all future years the staffing budget should be adjusted to conform to the results obtained from the patient classification system, *as per budgetary limits*. And, as unfair as it is, if budgetary constraints will not support the existing values in the patient classification system, then the latter system's values must be revised again, with the same relationship between patient classifications maintained, in order to conform to the budget. Otherwise, it will become a tool that cannot be used as intended.

Chapter 6
A Productivity and Cost Application

An Analytics Reporting Program

The purpose of the Productivity and Cost Application is to provide department directors and administrators with a performance summary for each department each payroll period throughout the year. The application evolved through the years to a data-based-driven application with each new download/input of payroll data, workload data, and non-salary expense data immediately available to each director and administrator on their desktop computers.

The data is presented consistently throughout the application, with the Payroll Periods displayed down the left-hand side of the screen in column A, the Actual data in column B, the Budget data in column C and the Variance in Column D. The bottom row contains the YTD data. Adjacent to the data table is a graph of the Actual and Budget data.

The Application consists of three modules. The first module (the default module) is the Budget module and consists of tabs for Paid Hours, Worked Hours, Overtime Hours, Paid FTEs, Worked FTEs, Salary Costs, Non-Salary Costs, Total Costs and Workload.

The second module is the Productivity module and consists of tabs for Paid Hours/Unit of Measure, Worked Hours/Unit of Measure and Total Costs/Unit of Measure. In addition, it indicates the expected Productivity, in terms of Paid Hours/Unit of Measure, for comparison with the Actual YTD achieved and the YTD totals that were budgeted. The expected productivity was derived by an independent source as a reference tool.

The third module is the Standard module and introduces a *flexible* budget approach versus the (absolute) budget illustrated in the first module, and consists of tabs for Paid Hours, Work Hours and Total Costs. On these screens, Standard Hours are used instead of Budget Hours and are calculated by multiplying the actual workload by the various productivity measures, e.g., the paid, worked and total costs times each parameter's budgeted productivity measure (Budgeted Paid Hours/Unit of Measure, Budgeted Work Hours/Unit of Measure and Budgeted Total Costs/Unit of Measure).

M. V. Calichman, *Essential Analytics for Hospital Managers*, SpringerBriefs in Health Care Management and Economics, https://doi.org/10.1007/978-3-030-16365-5_6

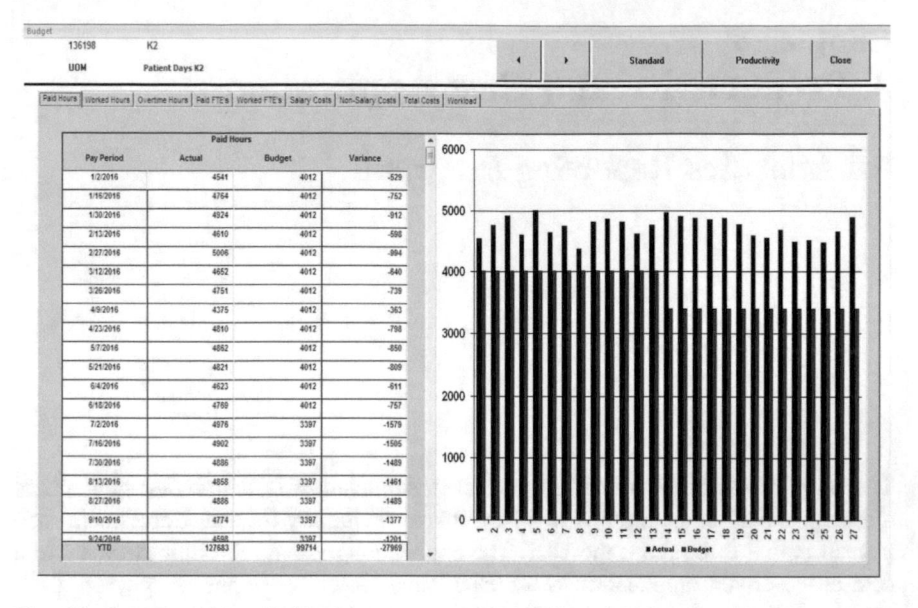

Fig. 6.1 Actual payroll period performance versus budget

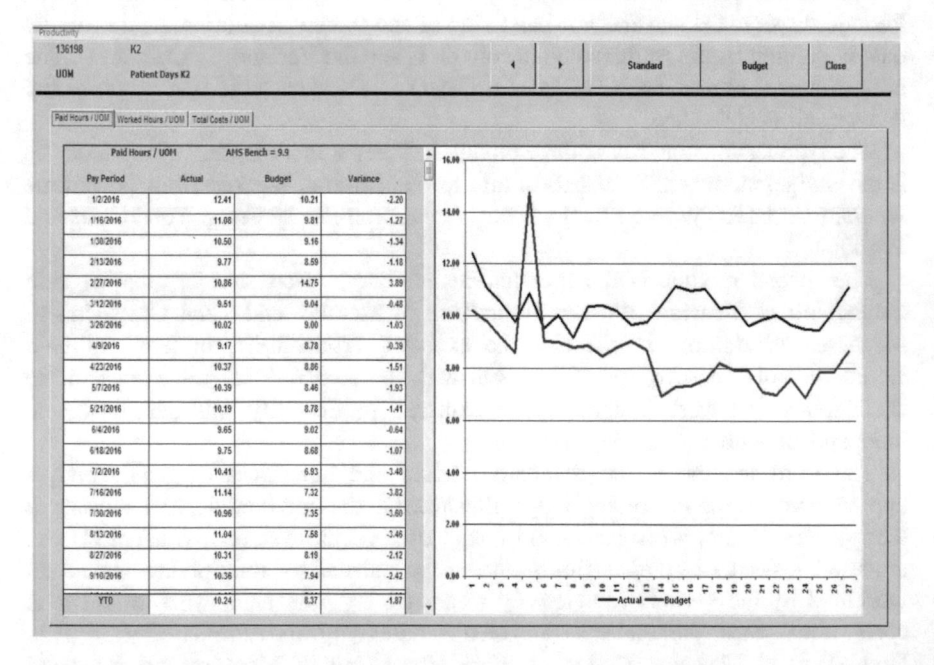

Fig. 6.2 Actual productivity versus budget productivity

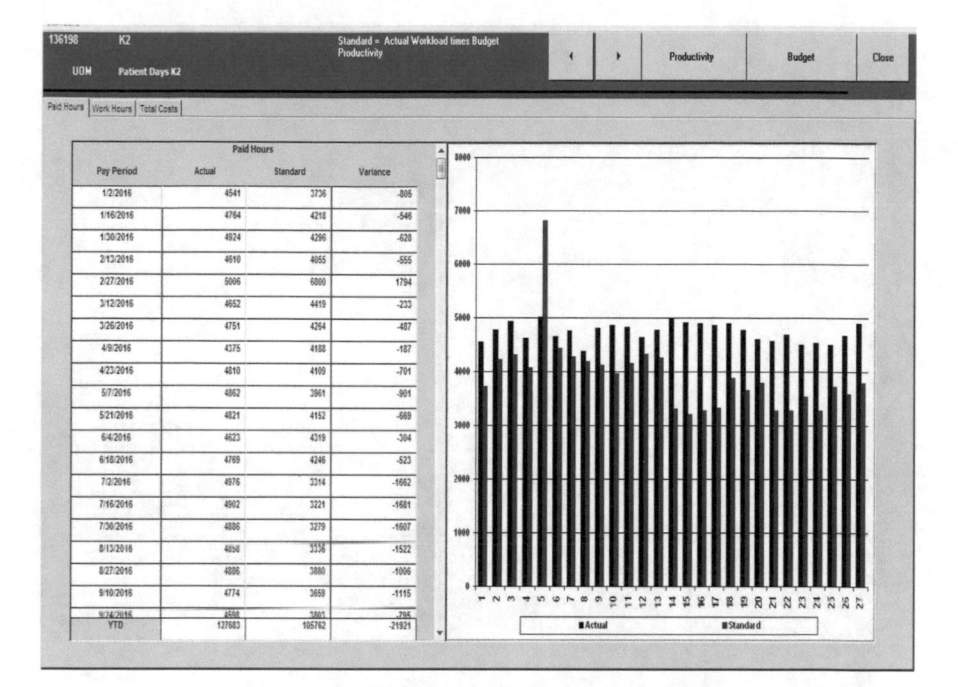

Fig. 6.3 Actual performance versus flexible budget (standard)

In short, if a floor nursing department was budgeted at 10.0 paid hours per patient day and was forecasted to service 30 patients per day, or 420 patients over each payroll period (30 × 14), the department would be budgeted (the absolute budget) for 56.0 fte's (420 × 10/75). However, if the department operated at its full capacity of 38 beds over the 14 day payroll period, or provided service to 532 patients (38 × 14), there would exist the rationale to increase (paid) staffing to 70.9 fte's (532 × 10/75); thus providing nursing with the justification for increasing staff during busy, but un-forecasted, spikes in workload.

Screen prints are included for the first screen in each module. See Figs. 6.1, 6.2 and 6.3.

Chapter 7
A CCU Bed Expansion

Poisson Arrival Rates and Summed Poisson with Truncated Data

Occasionally, there are problems when data is truncated, in part, by physical constraints preventing it from achieving a full and complete distribution. Consider the Intensive Care Unit (ICU) and the Critical Care Unit (CCU) whose beds are utilized to capacity almost every day. The hospital from which this data was drawn had a five-bed CCU unit and wanted to expand its number of beds. Table 7.1 illustrates a 180 day sample of patient days sorted by the number of days with 0–5 patients, while Fig. 7.1 illustrates a graph of that data. Note that the graph in Fig. 7.1 looks very much like the left side of a normal distribution curve that was discussed in Chaps. 3 and 4. Also note that with many more beds on this unit (let's say 15), its usage graph would most probably reflect a normal distribution, or bell curve.

From a review of the data, there were no days with zero patients and no days with only one patient. There were 4 days with two patients, 16 days with three patients and 44 days with four patients. And, finally, there were 116 days with five patients. That is all that was known. What is not known is how many of those 116 days with five patients would have had six, seven, eight or more patients had there been a sufficient number of beds. But what is a "sufficient" number of beds, or, how many beds, based upon the sample data, should this unit have?

If there were a normal distribution of data, or a close proximity of it, as Chaps. 3 and 4 indicate, we would plan for the required number of beds by adding 1.64 standard deviations to the average. But this would not be correct because the data does not approximate a normal distribution.

The data that does exist, however, permits one to determine the bed requirements at various confidence levels. A French mathematician, Semeon D. Poisson, described a distribution of data that can be utilized to solve this type of problem, now known as the "Poisson Distribution". Additional information on this distribution is available in most statistics textbooks.

As shown in Table 7.1, the data is arranged in columnar format, with columns for "Patient Occupancy", "Number of Days", "% (of Total Days)" and "Cumulative %" (the sum of the %'s to, and including, each row in the table). Now, in addition

© The Author(s), under exclusive license to Springer Nature Switzerland AG 2019 27
M. V. Calichman, *Essential Analytics for Hospital Managers*,
SpringerBriefs in Health Care Management and Economics,
https://doi.org/10.1007/978-3-030-16365-5_7

Table 7.1 Summary of CCU patient census (180 days sample period)

Patient occupancy	No. of days	%	Cumul. %	Actual mean = 4.5	% equal to or greater than known distribution statistics with mean equal to:		
					Mean = 4.0	Mean = 5.0	Mean = 6.0
0	0	0.0	0.0	100.0	100.0	100.0	100.0
1	0	0.0	0.0	100.0	98.2	99.3	99.7
2	4	2.2	2.2	100.0	90.8	96.0	98.2
3	16	8.9	11.1	97.8	76.2	87.5	93.8
4	44	24.4	35.6	89.0	56.7	73.5	84.8
5	116	64.4	100.0	64.6	37.1	56.0	71.4
6					21.5	38.4	55.4
7					11.1	23.8	39.3
8					5.1	11.3	25.6
9					2.1	6.8	15.2
10						3.2	8.3
11						1.4	4.2
12							2.0

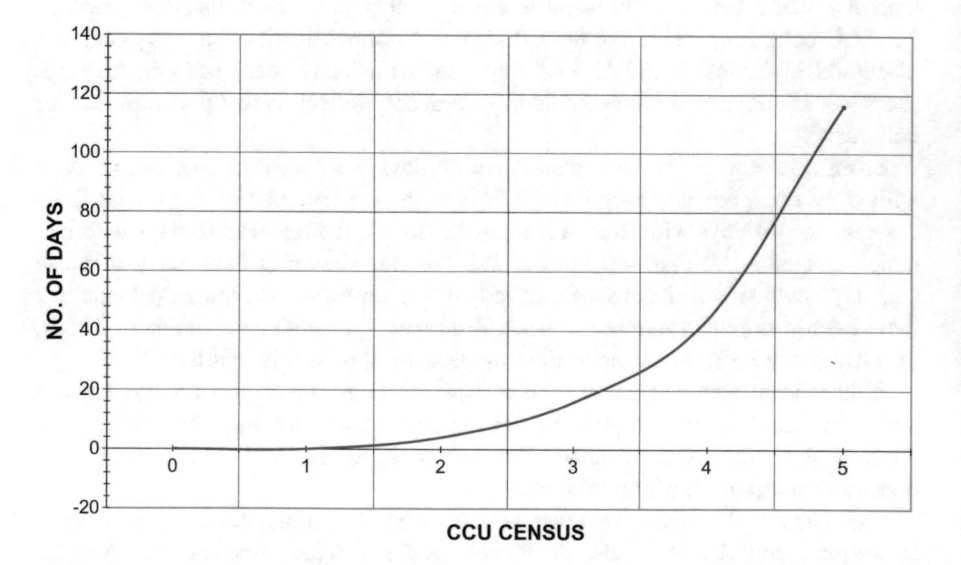

Fig. 7.1 Frequency distribution of CCU census

to knowing that there were three patients for 16 days, or 8.9% of the days in the sample period, one also knows that 11.1% of the time there were three or less patients (the number of days there were 3 patients, 2 patients, 1 patient or zero patients). In a similar manner, it can be stated that there were four or less patients 35.6% of the time and, of course, five or less patients 100% of the time.

Now add one more column entitled, "% Equal To or Greater Than" to change the format of the cumulative % so that the data can read, "% of the time the number of patients was equal to or greater than …". In this case, 97.8% of the time (100 − 2.2%) the number of patients was equal to or greater than three. 89.0% of the time the number of patients was equal to or greater than four and, lastly, 64.6% of the time the number of patients was equal to or greater than five. What is not known is the % of time the number of patients would have been equal to or greater than six, seven, eight or more patients.

However, with the data in this format, refer to a table of Summed Poisson, which may be found in any good statistics text book, extract the necessary data from appropriate means (in this case mean = 4, 5 and 6), arrange that data in a similar format and compare the extracted data, with means equal to 4, 5 and 6, with the actual data with mean equal to 4.5. Table 7.1 illustrates the completed table.

Figure 7.2 depicts a graph of the comparative statistics with mean equal to 4.5 (actual), 4.0, 5.0 and 6.0 and the number of patients from 0 to 12. Based upon the resulting graphs, select the best match of the data extracted from the table of Summed Poisson with the actual data. The graph of the actual data is extended

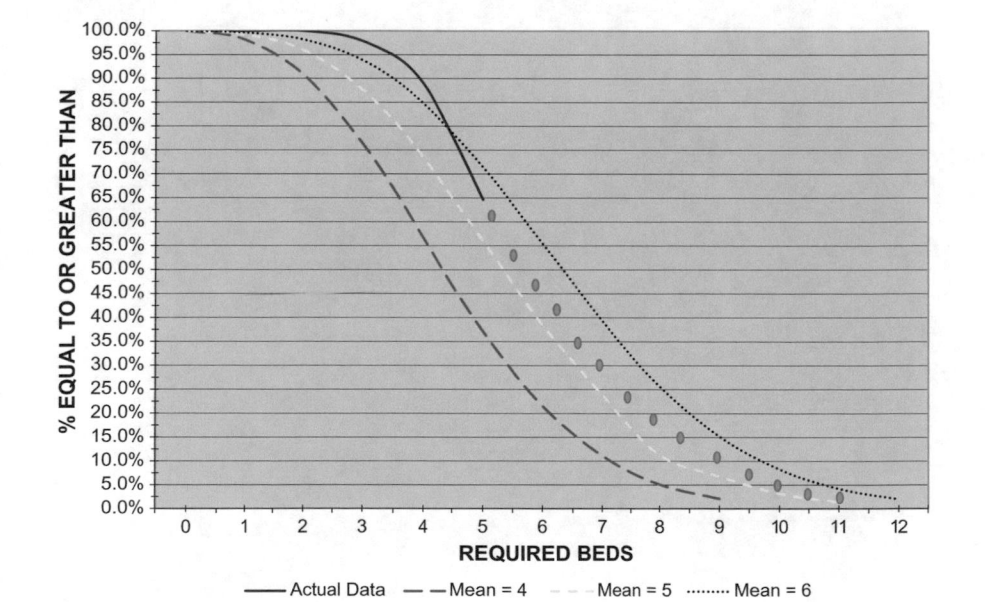

Fig. 7.2 Determination of CCU beds required

(free-hand), to approximate the graphs of the data extracted from the Summed Poisson table in a Statistics book.

It appears that the graph of actual data (when projected) will fall somewhere between data that is illustrated with means of 5.0 and 6.0 (or between 9 or 11 beds at the 95% confidence level). Thus, a mean of 5.5 patients appears to be a good approximation of what the actual mean might have been had it not been constrained by the maximum of five beds. Since the goal of this exercise would be to have a sufficient number of CCU beds for the patients requiring those beds at least 95% of the time, the conclusion was to expand the CCU unit to ten beds.

And from the table of the Summed Poisson with a mean equal to 5.5, the % of having 10 or less patients is equivalent to 0.0538, meaning that 10 CCU beds will suffice 94.62% of the time ($1.0000 - 0.0538 = 0.9462$).

Chapter 8
Inventory Management

The Minimization of the Total Costs of Ordering and Holding Supplies

In a previous chapter we discussed how to determine the appropriate quantity of inventory items to maintain on each nursing unit and other supply-user areas. That analysis was based solely upon usage statistics.

In this chapter we will discuss how to determine the appropriate item quantity to order from the vendor each time an order has to be placed—the *Economic Order Quantity*—as well as the determination as to when the order should be placed—the *Reorder Point*.

There are two basic costs associated with inventory; the cost of holding inventory (including the cost of money, pilferage, breakage, storage, obsolescence, etc.) and the cost of re-ordering (including the cost of inventorying, creating/conveying the order, receiving the order, processing the paperwork, etc.). The two costs are inversely related to one another. It is easily understood that somewhere between the following two options provides the optimum reordering schedule:

Option 1: Order the total year's requirements for medium TED stockings once each year

Option 2: Order 1/365 of the year's requirements each day

Obviously, Option 1 will have a maximum holding cost and a minimum re-order cost. Option 2, on the other hand, will have a minimum holding cost and a maximum re-order cost. But how are the optimum ordering schedule and minimum total cost derived?

Globally, holding cost is usually calculated at between 20 and 30% of product cost. Because in-hospital storage space is usually very expensive, the higher % is used for these calculations. Re-order cost varies, but is usually in the range of $6.00 per item (at the time this study was done … which was years ago). Figure 8.1 depicts a model that can be used to determine the EOQ, or economic order quantity.

Note the graph has "Quantity" along the "y" axis and "Time" along the "x" axis. Over time, the Quantity is used until it reaches zero. Sometime before the quantity reaches zero, it is re-ordered. This cycle repeats over time; quantity is used and reordered.

© The Author(s), under exclusive license to Springer Nature Switzerland AG 2019   31
M. V. Calichman, *Essential Analytics for Hospital Managers*,
SpringerBriefs in Health Care Management and Economics,
https://doi.org/10.1007/978-3-030-16365-5_8

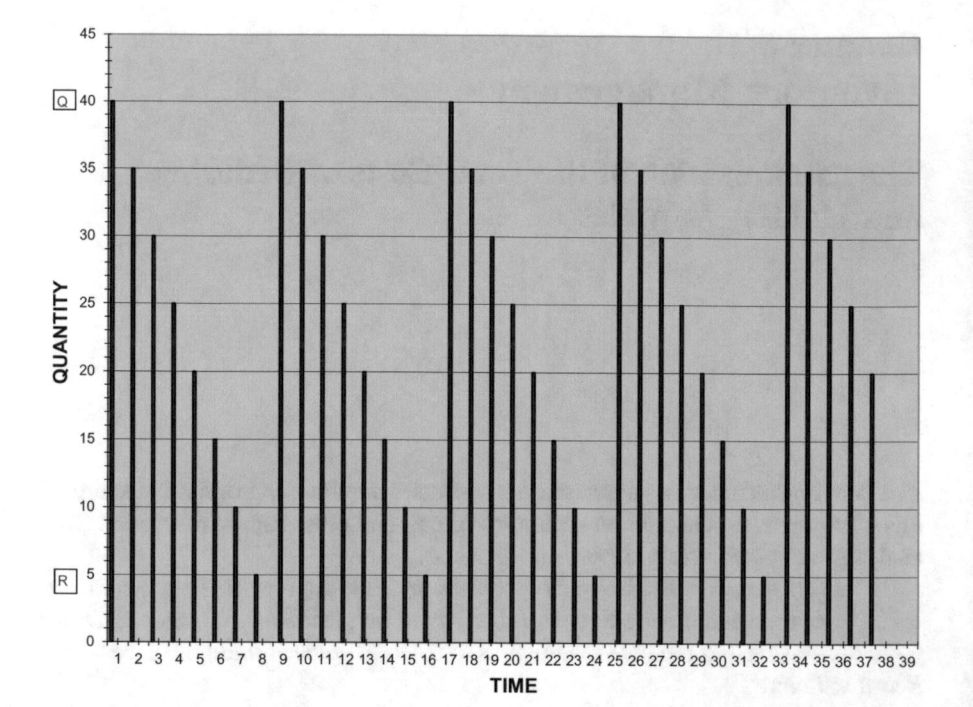

Fig. 8.1 Supply volume over time with constant usage

As mentioned before, the two basic costs associated with inventory is the holding cost and the re-order cost. The total cost equals the sum of the two, or:

$$\text{Total Cost} = \text{Holding Cost} + \text{Re−order Cost}$$
$$= (C_1Q)/2 + D(C_3)/Q \tag{8.1}$$

where

Q	Quantity to Order
Q/2	Average Quantity (range from 0 to Q)
D	Annual Demand
D/Q	No. of Orders/Year
C_1	Holding % times Price
C_3	Re-order Cost

From the calculus, take the first derivative of "Total Cost" with respect to time and solve for "Q" to minimize total costs, as follows:

$$\text{Minimum Cost} = (C_1)Q^0/2 - D(C_3)/Q^2 \qquad (8.2)$$

$$Q^2(C_1) = 2D(C_3)$$
$$Q = \text{Sqrt}(2DC_3)/(C_1)$$

This re-order Q is known as the EOQ or Economic Order Quantity. It tells us "how much" to order (Q) each time we re-order. Equally important is knowing "when" to order. Theoretically speaking, you re-order so that the new order comes in as the existing supply approaches zero. However, because demand and lead-time are not constant, it is prudent to build some additional stock in the system. In other words, the additional stock, most commonly referred to as "safety" stock, provides a buffer in case of increased demand, late delivery, etc. The mathematics for determining the reorder level is rather simple.

$$R = D/52(L + S) \qquad (8.3)$$

where

R   Re-order Point
D/52   Weekly Demand
L   Lead time (in weeks)
S   Safety stock (in weeks)

Let us now determine the Economic Order Quantity (Q) and the Re-order Point (R) for the following example:
Knee High Stockings

$$D = 2000 \text{ Dozen}$$
$$P = \$26.00 \text{ per Dozen}$$
$$C_1 = \$7.80 \ (30\% \times \$26.00)$$
$$C_3 = \$6.00$$
$$L = 1/2 \text{ Week}$$
$$S = 1/2 \text{ Week}$$

From above:
$$Q = \text{Sqrt}(2DC_3)/(C_1) \qquad (8.4)$$

And, $Q = \text{Sqrt}\,(2)(2000)(6)/(7.80)$
And, $Q = \text{Sqrt}\,(3076.923)$
And, $Q = 55.5$, or 56 dozen
And, also from above:

$$R = D/52(L+S)$$ (8.5)

And, $R = (2000)/52 \times (0.5 + 0.5)$
And, $R = 38.46 \times (1.0)$
And, $R = 38.46$, or 38 dozen

Thus, it is shown that to minimize the total inventory cost of knee high stockings, it would be prudent to order 56 dozen each time the on-hand supply decreases to 38 dozen. Because there should be approximately 19 dozen knee high stockings on hand each time a new order is received (the safety stock = ½ week's usage), inventory for this item should range between 19 dozen and 75 dozen (19 + 56), or an average inventory of 37.5 dozen, at a purchase cost of $975.

This item will be ordered approximately 36 times each year (2000/56), or approximately once every week and one half. Since items from this supplier may be ordered but once each week, it may be desirable to increase the safety stock from ½ week to one full week. This would have the impact of re-ordering when the on-hand is at 58 dozen [(2000)/52 × (0.5 + 1.0)] rather than at 38 dozen.

For proof that ordering 56 dozen knee high stockings each time the item is ordered minimizes inventory cycle costs, let's introduce other ordering patterns and compare total costs. Option 2 is to order a greater quantity each time the item is ordered while Option 3 is to order a less quantity each time the item is ordered.

Option 1:   Scientific Ordering using EOQ and R

$$
\begin{aligned}
\text{Total Cost} &= \text{Holding Cost} + \text{Re−order Cost} \\
&= (C_1 Q)/2 + (DC_3)/Q \\
&= (\$7.80 \times 56)/2 + (2000 \times \$6.00)/56 \\
&= \$218.40 + \$214.29 \\
&= \$432.69
\end{aligned}
$$ (8.6)

Option 2:   Order more—100 dozen each time the item is ordered

$$
\begin{aligned}
\text{Total Cost} &= (C_1 Q)/2 + (DC_3)/Q \\
&= (\$7.80 \times 100)/2 + (2000 \times \$6.00)/100 \\
&= \$390.00 + \$120.00 \\
&= \$510
\end{aligned}
$$ (8.7)

Option 3:   Order less—25 dozen each time the item is ordered

$$
\begin{aligned}
\text{Total Cost} &= (C_1 Q)/2 + (DC_3)/Q \\
&= (\$7.80 \times 25)/2 + (2000 \times \$6.00)/25 \\
&= \$97.50 + \$480.00 \\
&= \$577.50
\end{aligned}
\tag{8.8}
$$

The savings, as per the examples shown above, amounts to less than \$100 per year for the selected item. However, when consideration is given to the thousands of items ordered by an average hospital and the high cost of some of those supply items, especially for the O.R., other procedure rooms, the Pharmacy, etc., it is not uncommon for a 250–350 bed hospital to experience a one time savings in purchases in the range of \$200,000–\$500,000 when management follows this approach to ordering. Note the increase in the holding cost and the decrease in the re-order cost when a greater quantity is ordered and the decrease in the holding cost and the increase in the re-order cost when a less quantity is ordered each time the item is ordered. To minimize the total inventory cycle costs, use scientific ordering.

Chapter 9
Forecasting Resource Needs

Trend Lines and Seasonality

In this day and age, when graphical computer programs have become easier and easier to use, there is no need to discuss Trend Lines, etc. except for those readers who are curious as to how the trend line is derived and for those who are frustrated in their ability to show seasonality.

A number of years ago, a question arose as to the number of clinical examinations a particular hospital could expect to experience in the next few years, with the assumption that the same market forces would be in effect.

Although there are a number of statistical options available to project future occurrences, by far the most popular is *least squares*. It is defined by two algebraic expressions; the first stipulates that the sum of the variances between the actual data points and the trend line is zero. The second indicates that the sum of the squares of those variances is minimized (hence the name "least squares"). As Table 9.1 illustrates, the number of clinical examinations was recorded beginning January 1998 through September 2001, 45 months of data.

In order to derive the least squares trend line, it is necessary to determine the equation of that line. All straight lines have the form $y = a + bx$. The value of any "y" is determined by adding the value of where the line crosses the "y axis" (when $x = 0$) to the value of the slope of the line (b) times the value of x. Ordinarily, the first value of "x" would be 0 (the origin) and each subsequent value one greater than the previous value. Thus, it is expected that the "x" values would range from 0 to 44.

However, in order to make the mathematics simpler, it is beneficial to make the sum of all "x's" equal to zero. This is achieved by having the same number of negative points as positive points.

When there are an odd number of observations, as in this example with 45 data points, divide the number of data points by 2 and subtract 0.5 from the result to determine the number of both negative and positive values. Thus, in our example, there will be 22 negative and 22 positive values ($45/2 - 0.5$) and one zero value. Thus, the "x" values will range from −22 to +22. The sum of those values is

© The Author(s), under exclusive license to Springer Nature Switzerland AG 2019   37
M. V. Calichman, *Essential Analytics for Hospital Managers*,
SpringerBriefs in Health Care Management and Economics,
https://doi.org/10.1007/978-3-030-16365-5_9

Table 9.1 Forecasting clinical examinations—data

Date	x	y	xy	x^2
Jan 1998	−22	93	−2046	484
Feb	−21	83	−1743	441
Mar	−20	85	−1700	400
Apr	−19	85	−1615	361
May	−18	90	−1620	324
Jun	−17	93	−1581	289
Jul	−16	92	−1472	256
Aug	−15	87	−1305	225
Sep	−14	97	−1358	196
Oct	−13	163	−2119	169
Nov	−12	133	−1596	144
Dec	−11	93	−1023	121
Jan 1999	−10	68	−680	100
Feb	−9	92	−828	81
Mar	−8	101	−808	64
Apr	−7	97	−679	49
May	−6	93	−558	36
Jun	−5	79	−395	25
Jul	−4	68	−272	16
Aug	−3	44	−132	9
Sep	−2	78	−156	4
Oct	−1	126	−126	1
Nov	0	105	0	0
Dec	1	76	76	1
Jan 2000	2	102	204	4
Feb	3	85	255	9
Mar	4	111	444	16
Apr	5	120	600	25
May	6	129	774	36
Jun	7	85	595	49
Jul	8	90	720	64
Aug	9	84	756	81
Sep	10	91	910	100
Oct	11	131	1441	121
Nov	12	90	1080	144
Dec	13	94	1222	169
Jan 2001	14	109	1526	196
Feb	15	109	1635	225
Mar	16	144	2304	256
Apr	17	107	1819	289

(continued)

Table 9.1 (continued)

Date	x	y	xy	x^2
May	18	130	2340	324
Jun	19	116	2204	361
Jul	20	78	1560	400
Aug	21	73	1533	441
Sep	22	89	1958	484
Totals	0	4388	2144	7590

Formula for a straight line is:
y = a + bx

Normal equations for a straight line with sum of x = 0 are:
b = sum xy/sum x^2 (9.1)
a = sum y/n (9.2)
a = 4388/45
a = 97.51
b = 2144/7590
b = 0.28

At 11/99 trend line:
y_1 = 97.51 + 0.28x (9.3)

At 01/98 trend line:
y_0 = 91.35 + 0.28x (9.4)

obviously zero (With an even number of data points, divide the number of data points by 2 and subtract 0.5 from each point. If the illustrated example had one more month of data, or 46 data points, the range would advance from −22.5 to +22.5.).

Because it is necessary to derive the values for both "a' and "b" for the trend line, it is necessary to determine the sum of all "y" values, the sum of multiplying each "x" value by its corresponding "y" value and, lastly, the sum of the squares of each "x" value. Derive the "a" term by dividing the sum of the "y" values by the number of data points, or 4388 by 45. In a similar manner, derive the "b" term by dividing the sum of multiplying each "x" value by its corresponding "y" value by the sum of the squares of each "x" value, or 2144 by 7589. The corresponding "a" value is 97.51 and the "b" value is 0.28. Thus, the trend line, determined by the least squares method, is: y = 97.51 + 0.28x. Use Excel, or a similar spreadsheet, to perform the mathematics.

To move the origin back from November 1999 (the zero data point) to January 1998, simply multiply the existing value of "x" at that point (−22) by 0.28 to obtain −6.16 and subtract that amount from 97.51 to obtain 91.35. With this change, the trend line now becomes: y_0 = 91.35 + 0.28x. The slope of the line (the co-efficient of "x"), of course, remains unchanged.

In order to project the number of examinations in the future, use the formula of the trend line, with origin back at January 1998, and plug in 45 for October 2001, 46 for November 2001, 47 for December 2001, and so on and so forth. The number of examinations forecasted for each month, 2002–2005 is illustrated under "Trend" in Table 9.2. This is, of course, the unadjusted forecast. In order to account for monthly or seasonal fluctuations, it is necessary to continue making calculations (Table 9.3).

Table 9.2 Seasonality determination: actual versus trend

No.	Month	Actual	Trend	Adj. trend	No.	Month	Actual	Trend	Adj. trend
0	Jan-98	93	91		48	Jan-02		105	100
1	Feb	83	92		49	Feb		105	99
2	Mar	85	92		50	Mar		106	118
3	Apr	85	93		51	Apr		106	110
4	May	90	93		52	May		106	119
5	Jun	93	93		53	Jun		106	100
6	Jul	92	94		54	Jul		107	89
7	Aug	87	94		55	Aug		107	78
8	Sep	97	94		56	Sep		107	96
9	Oct	163	94		57	Oct		108	154
10	Nov	133	95		58	Nov		108	120
11	Dec	93	95		59	Dec		108	96
12	Jan-99	68	95		60	Jan-03		108	103
13	Feb	92	95		61	Feb		109	103
14	Mar	101	96		62	Mar		109	122
15	Apr	97	96		63	Apr		109	114
16	May	93	96		64	May		110	122
17	Jun	79	97		65	Jun		110	104
18	Jul	68	97		66	Jul		110	91
19	Aug	44	97		67	Aug		110	80
20	Sep	78	97		68	Sep		111	99
21	Oct	126	98		69	Oct		111	159
22	Nov	105	98		70	Nov		111	124
23	Dec	76	98		71	Dec		112	99
24	Jan-00	102	99		72	Jan-04		112	107
25	Feb	85	99		73	Feb		112	106
26	Mar	111	99		74	Mar		112	126
27	Apr	120	99		75	Apr		113	117
28	May	129	100		76	May		113	126
29	Jun	85	100		77	Jun		113	107
30	Jul	90	100		78	Jul		113	94
31	Aug	84	101		79	Aug		114	83
32	Sep	91	101		80	Sep		114	102

(continued)

Table 9.2 (continued)

No.	Month	Actual	Trend	Adj. trend	No.	Month	Actual	Trend	Adj. trend
33	Oct	131	101		81	Oct		114	164
34	Nov	90	101		82	Nov		115	128
35	Dec	94	102		83	Dec		115	102
36	Jan-01	109	102		84	Jan-05		115	110
37	Feb	109	102		85	Feb		115	109
38	Mar	144	102		86	Mar		116	130
39	Apr	107	103		87	Apr		116	121
40	May	130	103		88	May		116	130
41	Jun	116	103		89	Jun		117	110
42	Jul	78	104		90	Jul		117	97
43	Aug	73	104		91	Aug		117	85
44	Sep	89	104		92	Sep		117	105
45	Oct		104	149	93	Oct		118	168
46	Nov		105	117	94	Nov		118	132
47	Dec		105	93	95	Dec		118	105

First, it is necessary to return to the first month in order to determine the number of clinical exams that would have been projected each month using the trend line. January 1998 would be projected at 91.35, or 91 exams. Next, compare the actual results with the projected results each month. In as much as there were 93 exams actually provided by the clinic in January 1998 versus the 91 projected to have been provided, the actual is 102% of the projected when using the trend line. For the four years in the sample period, January averaged 96% of what the trend projected.

Because the sum of these averages is a little greater than 1200% (100% × 12 months), it is customary to divide 1200 by the actual sum of these monthly averages, 1212, to make certain that the sum of the monthly indexes adds to 1200. Thus, the adjusted seasonal index for January becomes 95 (96 × 1200/1212).

Lastly, apply the monthly indexes to the unadjusted forecast for each month to derive the adjusted forecast. For January 2002, the adjusted forecast is for 100 clinic visits. For a graphical comparison between the actual data, the computer generated trend line for the actual data, the unadjusted forecast line (same as trend line going forward) and the adjusted forecast line, refer to Fig. 9.1. You will well note the impact of seasonality on the data.

One last note, the individual analyst can decide not to use the arithmetic average as the "% of Trend" for each of the months. The analyst, based upon the data, might feel more comfortable and might eventually prove more accurate to use the median value or a modified mathematical average (by eliminating monthly highs and lows).

Table 9.3 Forecasting clinical examinations—summary

	Actual				From trend line				Actual as a % of trend line					
	1998	1999	2000	2001	1998	1999	2000	2001	1998	1999	2000	2001	Mean	Index
Jan	93	68	102	109	91	95	98	101	102	72	104	107	96	95
Feb	83	92	85	109	92	95	98	102	91	97	86	107	95	94
Mar	85	101	111	144	92	95	99	102	92	106	113	141	113	112
Apr	85	97	120	107	92	96	99	102	92	102	121	105	105	104
May	90	93	129	130	92	96	99	103	97	102	130	127	113	112
Jun	93	79	85	116	93	96	99	103	100	82	85	113	95	94
Jul	92	68	90	78	93	96	100	103	99	71	90	76	84	83
Aug	87	44	84	73	93	97	100	103	93	46	84	71	73	73
Sep	97	78	91	89	94	97	100	104	104	80	91	86	90	89
Oct	163	126	131		94	97	101		174	130	130		144	143
Nov	133	105	90		94	98	101		141	108	89		113	112
Dec	93	76	94		94	98	101		98	78	93		90	89
													1212	1200

	Unadjusted forecast				Adjusted forecast			
	2002	2003	2004	2005	2002	2003	2004	2005
Jan	105	108	112	115	100	103	107	110
Feb	105	109	112	115	99	103	106	109
Mar	106	109	112	116	118	122	126	130
Apr	106	109	113	116	110	114	117	121
May	106	110	113	116	119	122	126	130
Jun	106	110	113	117	100	104	107	110
Jul	107	110	113	117	89	91	94	97
Aug	107	110	114	117	78	80	83	85

(continued)

Table 9.3 (continued)

	Unadjusted forecast				Adjusted forecast			
	2002	2003	2004	2005	2002	2003	2004	2005
Sep	107	111	114	117	96	99	102	105
Oct	108	111	114	118	154	159	164	168
Nov	108	111	115	118	120	124	128	132
Dec	108	112	115	118	96	99	102	105

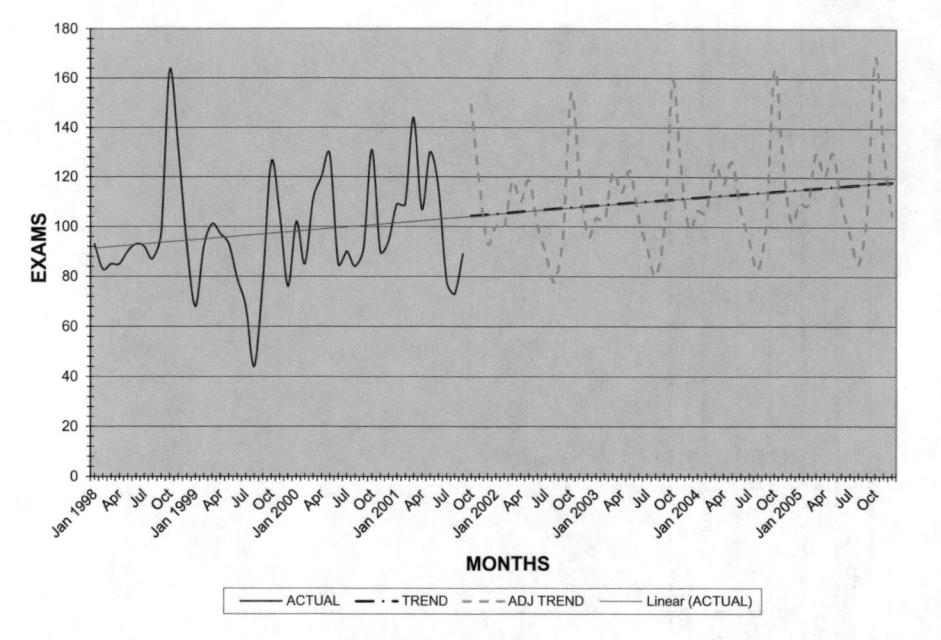

Fig. 9.1 Clinical examinations (actual and forecast)

Remember, however, that forecasting is predicated upon taking known data and projecting it forward. In the ensuing periods, market forces can change abruptly wreaking havoc to any and all calculations and causing the plans formulated from those calculations to become unacceptable.

Chapter 10
Determining an Emergency Department Trigger

Computer Simulation

Very often operational decisions are made without the benefit of appropriate data or the analysis of that data. Facilities are built that prove to be under-sized or over-sized simply because adequate tools were not available to the planners. Similarly, operations may be staffed either with too many providers or not enough providers, again because adequate tools are not available to the managers.

Computer simulation is one of those tools. Any flow of patients, materials, equipment, forms, etc. can be simulated and the resulting data analyzed. Computer simulation is an especially powerful tool for operations characterized by patient queues. I have used computer simulation in the last forty years, or so, to:

- Determine the number of Ambulatory Surgery Unit beds and Post Anesthesia Care Unit beds required at various volume levels for a planned Operating Room expansion.
- Determine the number of available stretchers required in the Emergency Department to reduce/maintain patient waiting times at manageable levels.
- Determine the number of nurses, as well as other health care professionals, required in a clinic to provide manageable patient waiting times.
- Determine the number of Environmental Services staff required in the O.R. to turnaround cases so as to minimize delays for surgeons, nurses and anesthesiologists.

Computer Simulation is not an exotic or an expensive management tool. Recent innovations have enabled simulations to be run on laptop computers. All that is required to simulate an operation on a computer are (1) appropriate knowledge of a computer program, (2) knowledge of the operational flow, and (3) pertinent operational statistics; e.g. what % of O.R. cases require 30 min. to perform, 60 min., 90 min., etc., what % of patients require 0 min. in PACU, 60 min. in PACU, 90 min. in PACU, etc., what % of ED patients require 60 min. throughput time, 90 min., 120 min., etc.

Using definitions and paraphrasing segments of James A. Chisman's fine book entitled, "Introduction to Simulation Modeling using GPSS/PC", the following can be stated:

M. V. Calichman, *Essential Analytics for Hospital Managers*,
SpringerBriefs in Health Care Management and Economics,
https://doi.org/10.1007/978-3-030-16365-5_10

A simulation model is a computer program representing the flow of people, material and/or information through a system, written for the purpose of experimenting off-line with alternate layouts, schedules, routing policies, equipment, input and service rates, storage space, work methods, etc. in order to find better ways of performing a task.

More specifically, simulation can help analyze and solve the type of problem associated with discrete waiting-lines (queues) such as those experienced in hospital emergency departments, recovery rooms (with an expected waiting time of zero), etc.

Queues or bottlenecks are caused by transactions (patients) arriving at a service area (E.D. providers) faster than the facility or person can service them.

The advantages of using a simulation model instead of experimenting with the real-world process itself are:

1. A simulation model of a system, because of the speed of computers, can run through several weeks, or months or even years of experience in a matter of minutes. Hence, several "What if?" scenarios can be evaluated within minutes or hours, whereas evaluating these scenarios on the real-world system could require months or years, at considerable cost, if it could be done at all.
2. Simulation modeling does not disrupt the operation of the real-world system. Hence, what might be considered "long-shot" or "harebrained" scenarios can be evaluated without bankrupting the hospital.
3. In a new design problem, you simply do not have a real-world process with which to experiment.

Assuming that the model is a true representation of the real system and that the data used is accurate, then a simulation model can save considerable time and money in designing a new system or in improving or fine-tuning an existing one.

As the title indicates, the purpose of this chapter is to determine when an administrative plan is to be deployed to discharge patients from floor beds, the so-called "trigger", in order to restore the appropriate number of available beds required in the Emergency Department (ED) to maintain ED patient waiting times at an acceptable level.

Throughout each day, as more and more ED beds (stretchers) become occupied by inpatients awaiting transfer to floor nursing units, subsequent patients are forced to wait longer and longer to be treated in the ED. One hospital introduced a plan of action to free beds on the nursing units by getting administration and physicians involved, whenever the "trigger" in the ED was reached. So, the question became, "how many beds in the ED are to be occupied by patients admitted to the hospital in order to trigger the administrative plan of action?"

On a 24 h basis, there appears to be an acceptable balance between patient arrivals in the ED and departures; 150 patients arrive during the day, 30 patients are admitted and transferred to appropriate beds on the nursing floors and the remaining 120 patients are treated and released. Unfortunately, arrivals are not consistent over the 24-h period. Approximately 2/3 of the patient load arrives between a twelve hour period; 11:00 a.m.–11:00 p.m. each day. During these hours the ED becomes inundated with patients, significantly delaying treatment times for the average ED patient.

Fig. 10.1 ED patient flow

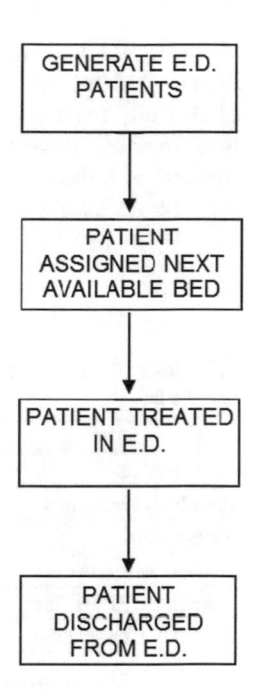

In an effort to determine the appropriate "trigger" for the Emergency Department to expedite the movement of admitted patients from the ED to the appropriate nursing units, two options were explored. The first option was to determine if there were any way to correlate overload conditions in the ED during the day with census data. No correlation was found.

The second option was to "simulate" ED throughput on a computer and to analyze under what conditions overload would result. This proved the better option. A simulation was programmed to obtain appropriate data from the arrival, treatment and discharge of ED patients during a 12-h peak period, during peak days, beginning with the arbitrary availability of 18 of the ED's 26 stretchers at the outset and then decrementing the number of available stretchers by one for each set of computer runs to determine when overload occurs. Appropriate distribution tables for patient generation times and patient treatment times were programmed. Using a random number generator, these time distribution tables and the appropriate flow of patients (see Fig. 10.1) will result in a reasonable and realistic simulation of patients flowing through the ED.

The simulations were designed to determine under what conditions the ED reaches overload. Although it was for hospital and ED management to make the final decision, the data seemed to indicate that overload is reached at the time the ED is reduced to only 13 of 26 available stretchers (with 13 stretchers occupied by ED patients admitted to the hospital but still in the ED).

Table 10.1 indicates the results of three runs with 18 available stretchers for 12 h in the ED. Although it was originally planned to simulate conditions for a minimum of 10 days, it was quite apparent, even with only three runs that the ED will not be in an overload situation with 18 available beds each day. A similar conclusion was reached with the ED having 15 available beds each day. When the ED runs with only 14 available beds all day, it is apparent that patients waiting to get treated in the ED lengthens.

Table 10.1 Results of ED trigger simulation 105 patients through ED (from 11:00 a.m. to 11:00 p.m.)

	18 available stretchers			15 available stretchers		
Run number	1	2	3	1	2	3
Available stretchers	**18**	**18**	**18**	**15**	**15**	**15**
Elapsed time	902	904	912	914	961	934
Patients generated	120	123	121	119	122	121
Maximum PTS waiting	4	2	1	7	5	4
Avg. no. PTS waiting	0	0	0	1	0	0
Average wait	0.7	0.1	0	6	0.8	3
0–15 min	117	122	121	97	118	112
15–30 min	3	0	0	14	1	7
30–45 min				8		
45–60 min						
60–75 min						
75–90 min						
90–105 min						
105–120 min						
120–135 min						
135–150 min						
150–165 min						
165–180 min						
180–195 min						
195–210 min						
210–225 min						
225–240 min						
240–255 min						
255–270 min						
270–285 min						
285–300 min						
300–315 min						
315–330 min						

(continued)

Table 10.1 (continued)

	18 available stretchers			15 available stretchers		
Run number	1	2	3	1	2	3
330–345 min						
345–360 min						
>360 min						

	14 available stretchers										
Run number	1	2	3	4	5	6	7	8	9	10	Avg.
Available stretchers	**14**	**14**	**14**	**14**	**14**	**14**	**14**	**14**	**14**	**14**	**14**
Elapsed time	927	957	923	949	926	944	970	924	908	915	**934**
Patients generated	123	122	128	131	138	143	125	134	145	127	**132**
Maximum PTS waiting	6	4	10	16	20	27	7	16	27	12	**15**
Avg. no. PTS waiting	2	0	4	4	7	11	1	4	12	5	**5**
Average wait	14	3	32	30	46	70	9	31	75	36	**35**
0–15 min	69	111	46	65	31	24	91	58	25	35	**56**
15–30 min	31	17	9	10	18	10	24	18	11	11	**16**
30–45 min	13		7	7	9	6	3	7	15	17	**8**
45–60 min	5		34	8	13	12		13	8	33	**13**
60–75 min			19	9	9	11		12	5	17	**8**
75–90 min			3	15	20	11		6	11	4	**7**
90–105 min				4		11		3	4	1	**2**
105–120 min						10		1	10		**2**
120–135 min						10			11		**2**
135–150 min						10			8		**2**
150–165 min						3			5		**1**
165–180 min									4		**0**
180–195 min									1		**0**
195–210 min											
210–225 min											
225–240 min											
240–255 min											
255–270 min											
270–285 min											
285–300 min											
300–315 min											
315–330 min											
330–345 min											
345–360 min											
>360 min											

(continued)

Table 10.1 (continued)

	13 available stretchers										
Run number	1	2	3	4	5	6	7	8	9	10	Avg.

	13 available stretchers										
Run number	1	2	3	4	5	6	7	8	9	10	Avg.
Available stretchers	**13**	**13**	**13**	**13**	**13**	**13**	**13**	**13**	**13**	**13**	**13**
Elapsed time	939	1018	1009	967	966	972	982	970	982	986	**979**
Patients generated	129	137	145	136	142	135	144	146	146	146	**140**
Maximum PTS waiting	18	25	21	28	19	25	21	29	29	30	**25**
Avg. no. PTS waiting	9	7	6	10	8	13	8	10	14	15	**10**
Average wait	67	54	45	67	58	88	62	65	94	103	**70**
0–15 min	18	53	39	28	16	14	30	25	14	16	**25**
15–30 min	10	3	18	3	1	11	2	5	1	5	**6**
30–45 min	15	6	4	11	23	5	7	15	12	5	**10**
45–60 min	8	9	12	19	20	11	12	29	10	6	**14**
60–75 min	3	4	17	16	16	2	14	13	7	11	**10**
75–90 min	8	6	19	10	15	6	12	6	16	8	**11**
90–105 min	27	13	**3**	11	22	7	22	3	6	14	**13**
105–120 min	22	5	3	8	4	18	12	2	9	3	**9**
120–135 min	6	9	**2**	**4**		14	6	**5**	22	5	**7**
135–150 min		9	**1**			14		**3**	5	8	**4**
150–165 min			**1**			13		**5**	1	11	**3**
165–180 min			3			1		4	8	14	**3**
180–195 min			**2**			1		2	5	5	**2**
195–210 min									**1**	5	**1**
210–225 min										**1**	0
225–240 min											
240–255 min											
255–270 min											
270–285 min											
285–300 min											
300–315 min											
315–330 min											
330–345 min											
345–360 min											
>360 min											

When the ED is down to 13 available beds, however, the situation worsens dramatically; with 63 of the 140 patients generated, on average, waiting greater than one hour for treatment (45% of all patients) and 20 of the 140 patients waiting in excess of 2 h (14% of all patients). Also, the maximum number of patients waiting increases to 25 patients (on average over the ten runs) with an average number of patients waiting at any one time of 10. Note that the simulation continues each run until 105 patients are terminated from the system.

Since significant waiting times begin to appear when the ED has only 13 of its 26 stretchers available for incoming patients, the administrative plan of action to free beds on the nursing units should be deployed before this situation is encountered, or when the ED has 14 of its stretchers available for incoming patients.

Chapter 11
Maximizing Net Revenue (Optimum Pricing)

Linear Programming

Hospitals may have greater than 12,000 individual fees in its charge master that are used to bill insurance companies and patients for services performed. Fees are more than likely increased across the board, usually on an annual basis, to keep pace with inflation and/or to derive additional income. A change in pricing strategy will enable hospitals to greatly increase its profitability. The strategy for achieving these results is to apply advanced mathematical techniques to maximize the amount of reimbursement received.

Most hospital revenue is based upon fixed reimbursement for services rendered. A much smaller portion is received from those provider plans that reimburse on a "% of Charges" basis. Thus, those charges that have a greater chance of being reimbursed as a % of Charges are those that the pricing model identifies for increase.

The key to the success of these pricing models is the Contribution Margin. It is defined as the amount of reimbursement received from the "% of Charges" plans divided by Gross Revenue (price × volume) from all plans. Thus, if charge item x has a price of $100 and is billed 1000 times during the year, the Gross Revenue for that item would be equal to $100,000. If the dollars received for charge item x from the % of Charges provider plans is $12,000 (Net Revenue), then the Contribution Margin for item x is 12% ($12,000/$100,000) and any subsequent price increase would return 12% of that increase in Net Revenue to the hospital.

What the Optimum Pricing Model does (by using a linear programming application) is to select those charges that make the greatest contribution to Net Revenue, increase (or decrease) the prices subject to administrative-directed constraints and begin what is an iterative process until maximum net revenue is achieved.

Examples of model constraints are as follows:

1. Maintain Gross Revenue at its existing level, increase it by 8%, or decrease it by 5%.

© The Author(s), under exclusive license to Springer Nature Switzerland AG 2019
53
M. V. Calichman, *Essential Analytics for Hospital Managers*,
SpringerBriefs in Health Care Management and Economics,
https://doi.org/10.1007/978-3-030-16365-5_11

2. Provide a range of price to cost ratios (for those hospitals with solid cost data) or price to (CMS) APC (for those hospitals with suspect or no cost data) from 2–10, 3–5, 5–10, or any such range.
3. Provide a range of new prices for those hospitals with no cost data (or no APC data) of from 75 to 125% of existing fees.
4. Cap any individual item increase at "x" dollars.

For one such project to derive Optimum Pricing, it was decided to set a pricing range from 2 times the APC cost to 10 times the APC cost. At first, any item price less than 2× its APC was increased to equal 2× its APC while any price greater than 10× its APC was lowered to 10× its APC.

This, in itself, created additional Net Revenue for the hospital as the impact from the increased prices far outweighed the impact from the price reductions. This approach, however, only analyzed those few charges that were outside the desired range. More importantly was to subject all of the prices to the model, including those that were already within the 2× to 10× range.

Thus, if a particular charge was sitting at 10 times its corresponding APC, but generated no Net Revenue, the model would reduce it to 2 times its APC without decreasing Net Revenue. This created significant opportunity for other APC related charges to increase (while maintaining a constant Gross Revenue). In a similar fashion, the model would price any charge that had a significant % of Net Revenue to Gross Revenue (and much volume) sitting at 2× its corresponding APC at or near 10× its APC.

In addition, it became apparent that the overwhelming majority of charges had no related APC. However, they all had a Contribution Margin, ranging from 0% to something higher. Thus, most of these charges (exceptions were made for Pharmacy charges and a few others) were able to be included in the model using the following constraints.

A. For APC related Charges

    a. A range of 2–7 times corresponding APCs
    b. No more than $500,000 in additional Net Revenue for any one charge
    c. A Gross Revenue no greater than 8% more than last year

B. For non-APC related Charges

    a. No more than 1.3 times the existing price
    b. No less than the existing price
    c. A Gross Revenue no greater than 8% more than last year

The new prices that satisfied these constraints added five times the amount to Net Revenue that was contributed by simply bringing all charges within the existing 2× to 10× range. And this was accomplished without any purchases of new equipment, without any new program initiatives and without any time delays.

See Table 11.1 (Sheets 1 and 2) for a look at a partial spreadsheet used to set-up the pricing model for one such hospital, spread over two pages. The table consists

Table 11.1 New pricing spread sheet

Sheet 1

Description of charge item	Total qty.	Total charges	Derived latest price 2014	Derived average price 2014	Derived contrib. margin	Anticipated 2014 gross revenue	Anticipated 2014 net revenue
PT MYCRD PRF SNG RST STRS	4	23,808	6369	6160	0.15750	49,283	7762
PT MYCRD PRF MLT RST STRS	1149	6,838,848	6369	6160	0.04167	14,156,415	589,912
PT MYCRD IMG MTBLC EVAL	24	119,952	5348	5173	0.02625	248,301	6518
PT MYCRD PRF SNG RST STRS	10	49,980	5348	5173	0.06300	103,459	6518
PT MYCRD PRF MLT RST STRS	76	379,848	5348	5173	0.02487	786,285	19,554
PTCT IMG SHULL/THGH STRS	94	469,812	5348	5173	0.04021	972,511	39,107
PTCT IMG WHOLE BODY	7	34,986	5348	5173	–	72,421	–
PTCT IMG SKULL THGH W MOD PS	41	204,918	5348	5173	0.04610	424,180	19,554
PTCT IMG WHOLE BODY WMOD PS	4	19,992	5348	5173	–	41,383	–
PTCT IMG WHOLE BODY PI	6	29,988	5348	5173	–	62,075	–
PTCT IMG SKULL THGH WMOD PI	26	99,840	4109	3974	0.02423	206,669	5008
Total charges	3,564,227	986,111,709				2,041,251,237	33,853,526

Sheet 2

(continued)

Table 11.1 (continued)

Sheet 2

Description of charge item	APC cost	Minimum multiple of APC	Maximum multiple of APC	Derived prices for 2015	Anticipated gross revenue 2015	Anticipated net revenue 2015	Increase in net revenue 2015–2014
PT MYCRD PRF SNG RST STRS	1569	3139	10,986	*10,986*	87,889	13,842	6080
PT MYCRD PRF MLT RST STRS	1569	3139	10,986	*10,986*	25,246,068	1,052,029	462,117
PT MYCRD IMG MTBLC EVAL	1569	3139	10,986	*10,986*	527,333	13,842	7325
PT MYCRD PRF SNG RST STRS	1569	3139	10,986	*10,986*	219,722	13,842	7325
PT MYCRD PRF MLT RST STRS	1569	3139	10,986	*10,986*	1,669,888	41,527	21,974
PTCT IMG SHULL/THGH	1569	3139	10,986	*10,986*	2,065,388	83,055	43,948
PTCT IMG WHOLE BODY	1569	3139	10,986	*3139*	43,944	–	–
PTCT IMG SKULL THGH W MOD PS	1569	3139	10,986	*10,986*	900,861	41,527	21,974
PTCT IMG WHOLE BODY WMOD PS	1569	3139	10,986	*3139*	25,111	–	–
PTCT IMG WHOLE BODY PI	1569	3139	10,986	*3139*	37,667	–	–
PTCT IMG SKULL THGH WMOD PI	1569	3139	10,986	*10,986*	571,277	13,842	8835
Total charges					2,116,003,433	42,754,304	8,900,778

Italic values denotes the results of the application

of 11 of the more than 4000 charge items considered for a pricing change. The first sheet indicates the following:

A. Description of Charge Item: A brief description of the charge.
B. Total Quantity: The number of times the charge item was generated over the stipulated period of time. There were over 3.5 million charge items generated during the six month period.
C. Total Charges: The total dollar charges for each item listed. These charges amounted to almost $1 billion in total in the six month period.
D. Derived Latest Price: The price the last time the price was updated.
E. Derived Average Price: The average price charged over the stipulated period of time ($ charges divided by number of units generated).
F. Derived Contribution Margin: The % of Gross Revenue contributed for each charge item by those insurance companies and/or plans that reimburse on a "% of Charges" basis. The contribution margins are derived on a separate worksheet.
G. Anticipated 2014 Gross Revenue: Each charge item's number of times generated for the full year × the item's average price over the year, summed for a total of greater than $2 billion.
H. Anticipated 2014 Net Revenue: Each item's Gross Revenue × the item's Contribution Margin. For the base year, the net revenue for the 4000+ charge items amounted to almost $34 million.

The second sheet indicates the following:

I. APC: The Medicare APC Cost to be obtained from the hospital (costs from the hospital's cost accounting system could be used instead).
J. Minimum Multiple of APC: For this hospital, the minimum revised price was set at 2× the APC Cost.
K. Maximum Multiple of APC: For this hospital, the maximum revised price was set at 7× the APC Cost.
L. Derived Price for 2015: The price determined by the linear programming application to satisfy all given constraints, e.g. required price range, no greater than the stipulated new gross revenue, etc. As indicated the primary constraint was that each item's new price would be in a range of 2–7× its APC value (The results in this column are derived by the application after all of the data are entered).
M. Anticipated 2015 Gross Revenue: The item's 2014 quantity × the item's new price. Assuming no significant change in each item's following year volume and a very similar profile of patient reimbursement results in an anticipated 2015 Gross Revenue of slightly greater than $2.1 billion.
N. Anticipated 2015 Net Revenue: Each item's 2015 Gross Revenue × the item's Contribution Margin, summed for the year will approximate $42.8 million
O. Increase in Net Revenue: The Anticipated 2015 Net Revenue less the 2014 Net Revenue, or almost $9 million. Note that the % of net revenue to gross revenue will increase from 1.66 to 2.02% by having the application scientifically select

which charge items to increase and by how much over the range of 2–7× their respective APC value.

The lesson learned is that an across-the-board percent increase in prices each year is to be discouraged, as it continues to inflate gross revenue and has a less than desired impact on net revenue. Hospital profitability (net revenue) can be significantly increased by setting up an Optimum Pricing Model, subject to administrative-approved constraints. Millions of dollars are at stake.

Chapter 12
Eliminating Hospital Overcrowding (Optimum O.R. Scheduling)

Linear Programming

Hospitals continue to allocate O.R. and other procedural cases using available time in the O.R.s and procedural labs as the one and only constraint to scheduling cases. In addition, they allow each surgeon to decide the days they prefer to operate, which is usually limited to Monday–Friday; allowing expensive resources to remain dormant and unused almost 30% of the time each week. Is it any wonder, therefore, that this scheduling system results in an inconsistent need for hospital beds during the week; increasing the demand for beds Monday through Thursday and then decreasing the demand from Friday through Sunday; as opposed to a scheduling system that provides maximum use of beds seven days each week, with no overcrowding?

What is more problematic is that there appears to be no awareness that a full (surgical and procedural) census could be achieved each day of the week by developing scheduling algorithms that take into consideration each discipline's demand for service, its unique distribution of lengths of stay, its Operating Room (O.R.) and Laboratory (Lab) time requirements, and all other operating constraints; as well as the hospital's overall bed capacity, O.R. and Lab scheduling capacities, etc.

The object of this scheduling application (the objective function in the linear programming models) is to maximize the total number of cases performed each week, whether it be over a five day or six day (preferred) period. Note that a seventh operational day is not necessary as model after model indicate that 100% of beds can be filled each day, each week with a six-day schedule.

For every hospital there exists a unique O.R. schedule that will:

(a) Ensure smooth patient flow to, through and from the O.R.
(b) That will enable the hospital to comply with all of its operating constraints.

Based on Calichman (2005). Reprinted with the permission of the copyright owner.

(c) That will balance the demand for surgical beds with the supply of surgical beds each day of the week (one of the primary constraints) and, in essence, eliminate the fluctuation in patient days during the week that leads to days of patient overcrowding.

The methodology employed to achieve this engineered patient schedule is, "Optimum O.R. Scheduling".

Optimum O.R. Scheduling was developed in the late 70s. A hospital administrator was seeking an O.R. schedule that would eliminate the need to cancel surgical procedures each week because of an insufficient number of surgical beds available (unoccupied) on the day the patients were scheduled to arrive at the hospital (usually the day before surgery). It became apparent, after reviewing the hospital's bed utilization statistics, that an O.R. schedule eliminating cancellations was possible, as there were a greater number of beds unoccupied during the week than those cancelled patients (eighteen per week, on average) would have required had they been operated upon. Operations simply had to be scheduled on different days to minimize and balance the number of floor beds required each day.

The bed utilization review indicated a hospital fully occupied from Sunday night through Thursday night, but significantly underutilized both Friday and Saturday nights. The challenge, therefore, was to design a schedule that would utilize each and every surgical bed in the hospital seven days each week; neither one bed more nor one bed less. Thus, there was an over-riding need to relate O.R. cases to bed use. In other words, the demand for beds had to be adjusted (by scheduling) so that it was consistent with the supply of those beds, seven days each week (Note: If hospitals elect not to run O.R. schedules on Saturday (or Sunday), there will be days during the week that surgical beds lie vacant, even under this optimization approach. Also, in those hospitals without designated medical or surgical beds, the methodology will determine the number of surgical beds required to bring it in balance with the O.R. activity, be it a 5 or 6-day per week schedule.).

Although all of the data is important, determining the distribution of lengths of stay for each surgical category and being able to utilize that data is the essential step in deriving the Optimum O.R. Schedule. The derived O.R. schedule for that initial installation enabled the hospital to eliminate its 18 cancellations each week and increase its revenue by 3%, by virtue of gaining increased utilization of its beds (Note: As some of you may not be aware, when *Optimum Scheduling* was first utilized, hospitals were reimbursed on a per diem basis.).

Unfortunately, soon after that implementation there was a *sea change* in the greater hospital environment. Two events of seismic proportion occurred that discouraged a widespread distribution of this application at other facilities. The first event was the change in reimbursement from a per diem rate to a DRG rate. The second was the change from inpatient reimbursement to outpatient reimbursement for many procedures. These dual events, occurring within months of each other, significantly reduced overall census statistics at most hospitals. Hospitals that had utilization statistics in the mid-to-high nineties became hospitals with utilization statistics in the high seventies to low eighties. These events seemingly obviated the need for Optimum O.R. Scheduling. With reduced average length of stay and reduced inpatient surgical volume, surgical beds no longer were an operational constraint.

However, many years later, another hospital requested a review of its O.R. schedule. The review indicated that little had changed. Hospitals were still developing daily O.R. schedules based upon one parameter—O.R. time—and were still experiencing an imbalance between the supply of, and the demand for, their surgical beds.

What did change, however, was how the imbalance was manifested. No longer were surgical cases cancelled the day before surgery. Now patients, for the most part, were arriving in the early morning on the day of surgery (without any consideration given to the availability of beds throughout the day). On the occasions when all floor beds are occupied (a continuing problem in some facilities), post-surgical patients find themselves "recovering" in the PACU for extended hours, if not days.

Other surgical patients may experience gridlock in the O.R.—not being able to flow into the PACU after surgery because beds aren't available in that location either; and still other patients are, indeed, cancelled (to a much lesser extent than what previously occurred). Since floor beds aren't available to post-surgical patients, neither are they available to incoming Emergency Department patients that are admitted to the hospital. These patients might end up spending days in the ED. These situations can all be changed with *Optimum O.R. Scheduling*.

In order to derive an Optimum O.R. Schedule for a facility, it is important to gather the appropriate data, use the data to relate the unknowns to the facility's operational constraints, determine the metric to be maximized (usually total cases), set-up the scheduling "problem" on a computerized spread sheet and solve the problem (Technically, since the solution requires a whole integer solution, i.e., 6 inpatient General Surgery cases on a Wednesday and not 6.2 cases, software will have to be purchased. The application embedded in Excel (Solver) does not have the capacity to return a whole integer solution.).

It would be best to follow these sequential steps:

1. Categorize all surgical patients (or all hospital patients) into groups or sub-groups of patients that hospital personnel are familiar with—see Table 12.1 for one such simplified grouping.
2. Create an unknown, X_i, for each patient category for each day of the week. In Table 12.1, for example, X_{29} is the number of (as yet unknown) inpatient ENT cases that will be scheduled for Mondays, X_{54} the number of (as yet unknown) outpatient AICD and Pacers cases that will be scheduled for Fridays, etc. The computer will determine the value for each of the 189 unknowns, which are the number of cases to perform by category each day of the week.
3. Obtain all necessary and appropriate data and relate the unknowns to that data.

    a. Determine the average weekly demand for each category and make that the minimum number to be scheduled each week.

Table 12.1 Problem set-up

Surgical categories	Avg. case mins (incl. TRNRND)	M	T	W	TH	F	S	SU	Weekly totals Avg. weekly vol.	Max. weekly vol.	Min. cases per day M–F S–Su	Min. cases per day (EMRG) S–Su	Max. cases per day M–Sun
OHS—INPT	361	X_1	X_2	X_3	X_4	X_5	X_6	X_7	20	20	1	8	8
TAVR (O.R.)—INPT	240	X_8	X_9	X_{10}	X_{11}	X_{12}	X_{13}	X_{14}	6	6	0	3	3
Bariatrics—INPT	90	X_{15}	X_{16}	X_{17}	X_{18}	X_{19}	X_{20}	X_{21}	2	2	0	4	4
Cardiology—INPT	0	X_{22}	X_{23}	X_{24}	X_{25}	X_{26}	X_{27}	X_{28}	0	0	0	1	1
ENT—INPT	95	X_{29}	X_{30}	X_{31}	X_{32}	X_{33}	X_{34}	X_{35}	2	2	0	2	2
ENT—OPT	110	X_{36}	X_{37}	X_{38}	X_{39}	X_{40}	X_{41}	X_{42}	7	7	0	4	4
AICD and PACERS—INPT	150	X_{43}	X_{44}	X_{45}	X_{46}	X_{47}	X_{48}	X_{49}	17	17	2	10	10
AICD and PACERS—OPT	125	X_{50}	X_{51}	X_{52}	X_{53}	X_{54}	X_{55}	X_{56}	16	16	0	6	6
Eye—OPT	55	X_{57}	X_{58}	X_{59}	X_{60}	X_{61}	X_{62}	X_{63}	13	13	0	10	10
General—INPT	235	X_{64}	X_{65}	X_{66}	X_{67}	X_{68}	X_{69}	X_{70}	22	22	2	10	10
General—OPT	140	X_{71}	X_{72}	X_{73}	X_{74}	X_{75}	X_{76}	X_{77}	31	31	0	10	10
GU—INPT	195	X_{78}	X_{79}	X_{80}	X_{81}	X_{82}	X_{83}	X_{84}	7	7	0	5	5
GU—OPT	110	X_{85}	X_{86}	X_{87}	X_{88}	X_{89}	X_{90}	X_{91}	14	14	0	4	4
Neuro/spine—NPT	250	X_{92}	X_{93}	X_{94}	X_{95}	X_{96}	X_{97}	X_{98}	8	8	0	3	3
Neuro/spine—OPT	160	X_{99}	X_{100}	X_{101}	X_{102}	X_{103}	X_{104}	X_{105}	2	2	0	1	1
Ortho—joints—INPT	190	X_{106}	X_{107}	X_{108}	X_{109}	X_{110}	X_{111}	X_{112}	23	23	2	8	8
Ortho—other—INPT	190	X_{113}	X_{114}	X_{115}	X_{116}	X_{117}	X_{118}	X_{119}	7	7	1	3	3
Orthopedics—OPT	120	X_{120}	X_{121}	X_{122}	X_{123}	X_{124}	X_{125}	X_{126}	28	28	3	15	15
Plastic—INPT	240	X_{127}	X_{128}	X_{129}	X_{130}	X_{131}	X_{132}	X_{133}	1	1	0	3	3
Plastic—OPT	165	X_{134}	X_{135}	X_{136}	X_{137}	X_{138}	X_{139}	X_{140}	3	3	0	3	3

(continued)

Table 12.1 (continued)

Surgical categories	Avg. case mins (incl. TRNRND)	M	T	W	TH	F	S	SU	Weekly totals		Min. cases per day M-F	Min. cases per day (EMRG) S-Su	Max. cases per day M-Sun
									Avg. weekly vol.	Max. weekly vol.			
PMT—OPT	35	X_{141}	X_{142}	X_{143}	X_{144}	X_{145}	X_{146}	X_{147}	32	32	0		15
Thoracic—INPT	165	X_{148}	X_{149}	X_{150}	X_{151}	X_{152}	X_{153}	X_{154}	6	6	0		4
Thoracic—OPT	120	X_{155}	X_{156}	X_{157}	X_{158}	X_{159}	X_{160}	X_{161}	2	2	0		2
Vascular—INPT	155	X_{162}	X_{163}	X_{164}	X_{165}	X_{166}	X_{167}	X_{168}	33	33	3		12
Vascular—OPT	125	X_{169}	X_{170}	X_{171}	X_{172}	X_{173}	X_{174}	X_{175}	16	16	0		5
All other—INPT	90	X_{176}	X_{177}	X_{178}	X_{179}	X_{180}	X_{181}	X_{182}	2	2	1		1
TAVR (CATH lab)—INPT	240	X_{183}	X_{184}	X_{185}	X_{186}	X_{187}	X_{188}	X_{189}	4	4	0		2
Total cases									324	324			

b. Determine the maximum number of cases that could possibly be scheduled each week for each category. These limits will provide both the minimum and the maximum weekly demand constraints. In the calculus of the problem, for example, the sum of X_{78}–X_{84}, the number of inpatient General Urology cases to schedule each week, must be equal to, or greater than 7 and equal to, or less than 7. This, of course, will return a total of 7 cases. If, of course, the hospital were looking for the computer to make the determination of Inpatient General Urology cases to perform each week, subject to constraints, an expanded range would have been specified (7–10, perhaps).

c. Determine if there are daily, restrictions for reasons of equipment, staff availability, etc., that have to be considered. We could stipulate that X_{36}–X_{42} must be less than 2 each day, for example. If the hospital does not schedule surgery on Saturdays, then the unknowns for that day, e.g., X_6, X_{13}, X_{20}, etc. would be set equal to zero. In the example illustrated, it was indicated that there must be at least two inpatient General Surgery cases each day and one inpatient Orthopedics—Other case each day, M–F, and that no cases would be scheduled for the weekends.

d. Determine the O.R. time, including turnaround time required for the average case in each category. Note the average elapsed O.R. times in Table 12.1. Also determine the amount of O.R. time available each day (see Table 12.6 —last row of data). The number of cases scheduled each day multiplied by the applicable average time to perform each case (including turnaround) must be equal to or less than the total O.R. time available. And inasmuch as 100% utilization of O.R. time, even when turnaround time is included, is impractical, multiply the O.R. time available each day by 87.5% (7/8) to derive a more practical constraint. Also, since this is a relatively "soft" constraint, this number can be adjusted upwards, if, and as, required.

e. Obtain from Information Systems, a listing of all patients operated upon during the last 6–12 months with sufficient information (appropriate group, date admitted, date of operation, date discharged, etc.) so that the % of (Adult) patients operated upon requiring beds each day, by category, can be determined—see Table 12.2. For example, if 100% of ENT inpatients require a bed on the day of surgery (day Op), and 5% of those patients are still in the hospital 7 days later (day 8) and 0.9% of the patients are still in the hospital 14 days later (day 15), then beds must be supplied for 105.9% of ENT inpatients on the specific days of the week those patients are operated upon … plus the appropriate number of beds required for patients operated upon the other six days of the week. Note that Table 12.2 illustrates data for a 21-day length-of-stay period.

f. Obtain from the Nursing department, the number of beds available each day for surgical patients, or by category, if required. The number of patients requiring beds each day must be less than or equal to that number (see Table 12.4). For the illustrated example, all surgical patients each day must be less than the 127 beds available. Please note that if the O.R. Schedule were to be spread over a six day period, rather than the M–F period used,

Table 12.2 % of patients requiring beds by grouped days of recovery LOS window

Surgical categories	−1 + 6 + 13 (%)	Op, +7, +14 (%)	+1, +8, +15 (%)	+2, +9, +16 (%)	+3, +10, +17 (%)	+4, +11, +18 (%)	+5, +12, +19 (%)	Beds req'd for Pt cases (%)	LOS (max'd @ 21 days)
OHS—INPT	144.1	152.6	140.8	133.6	128.4	118.0	96.7	914	9.14
TAVR (O.R.)—INPT	64.3	120.9	115.7	67.8	56.5	45.2	31.3	502	5.02
Bariatrics—INPT	0.0	100.0	0.0	0.0	0.0	0.0	0.0	100	1.00
Cardiology—INPT	79.6	109.7	58.1	41.9	36.6	20.4	14.0	360	3.60
ENT—INPT	35.3	105.9	23.5	23.5	17.6	17.6	11.8	235	2.35
ENT—OPT								0.00	0.00
AICD and PACERS —INPT	105.6	109.9	63.0	42.0	32.7	22.8	20.4	396	3.96
AICD and PACERS —OPT								0.00	0.00
Eye—OPT								0.00	0.00
General—INPT	56.5	122.0	92.8	73.5	58.3	44.4	39.5	487	4.87
General—OPT								0.00	0.00
GU—INPT	55.0	110.0	76.7	65.0	58.3	43.3	35.0	443	4.43
GU—OPT								0.00	0.00
Neuro/spine—INPT	93.8	132.3	113.8	106.2	87.7	73.8	61.5	669	6.69
Neuro/spine—OPT								0.00	0.00
Ortho—joints—INPT	9.1	103.0	98.3	77.2	24.6	12.5	7.3	332	3.32
Ortho—other—INPT	83.1	121.7	109.6	88.0	59.0	45.3	38.6	546	5.46
Orthopedics—OPT								0.00	0.00
Plastic—INPT	90.5	133.3	119.0	85.7	71.4	61.9	57.1	619	6.19
Plastic—OPT								0.00	0.00
PMT—OPT								0.00	0.00

(continued)

Table 12.2 (continued)

Surgical categories	−1 + 6 + 13 (%)	Op, +7, +14 (%)	+1, +8, +15 (%)	+2, +9, +16 (%)	+3, +10, +17 (%)	+4, +11, +18 (%)	+5, +12, +19 (%)	Beds req'd for Pt cases (%)	LOS (max'd @ 21 days)
Thoracic—INPT	115.6	146.7	126.7	122.2	111.1	93.3	68.9	784	7.84
Thoracic—OPT									0.00
Vascular—INPT	98.1	128.4	90.5	70.1	62.6	50.7	46.4	547	5.47
Vascular—OPT									0.00
All other—INPT	76.3	116.9	88.1	67.8	50.8	33.9	28.8	463	4.63
TAVR (CATH lab)—INPT	64.3	120.9	115.7	67.8	56.5	45.2	31.3	502	5.02

there would be a need for approximately 123 surgical beds, a slight decrease. Note how each day, Monday–Friday, there is a consistent need for 127 beds. A balanced need could be extended over the seven day period, if the hospital had opted for a six day per week schedule.

g. With administrative approval, allow for surgeon preference. For example, it might prove practical to assign X_4 a minimum value of 2 in as much as "Dr. Smith", an OHS surgeon, can only provide service to the hospital on Thursdays (It makes sense, however, to derive the solution both ways, with the constraint and without the constraint to provide administration with the cost [usually in the need for additional beds] of allowing Dr. Smith to operate on Thursdays only.).

4. Enter the appropriate formulae into the appropriate cells in both Excel and the embedded (Linear Programming) Solver to obtain the optimum O.R. schedule for the hospital (Note that other spreadsheets also have optimization features similar to Solver.).

5. Solve the scheduling problem and validate the results (see Table 12.3 for the results and Tables 12.4, 12.5, 12.6 and 12.7 for the impact of the derived schedule on the constraints).

Unlike the late 70s when special software applications had to be purchased to solve this type of problem, it can now be solved (to decimal equivalent results) utilizing an embedded feature of Excel appropriately named the Solver. Excel and the Solver module both require knowledge in how they work. Although the use of neither application is within the scope of this paper, suffice to say that the mastery of the applications is well within the capabilities of personnel employed by the hospital.

The Solver is a linear programming engine (an optimizer) that will proceed through algorithms (solution cycles) until it maximizes the objective function—the parameter chosen for maximization (number of cases, for example) while satisfying all hospital constraints, providing there exists at least one feasible solution to the "problem"—that the problem can, indeed, be solved. Because the Solver application embedded in Excel, in all probability, will not be able to provide an "integer" solution, an enhanced version will have to be purchased.

Table 12.3 illustrates the resulting Optimum O.R. Schedule for the sample hospital and its scheduled number of cases. Note that the weekly schedule for 324 total cases is what is required, and, as is illustrated, achieved.

It is important to validate the results by making certain that the derived schedule complies with all operational constraints. As Table 12.4 indicates, the primary constraint, surgical beds, is complied with. In total, there were 127 beds used as the constraint. All 127 beds will be fully utilized on Monday–Friday and less utilized Saturday and Sunday. This illustrates how difficult it is to gain full utilization of beds without operating on a sixth day.

Table 12.5 indicates the impact the schedule has on demand by service. Note that each patient category's weekly totals, as well as daily totals, are within the

Table 12.3 Optimum weekly schedule

	M	T	W	TH	F	S	SU	Total	Min/day	Max/day
								324 = total cases		
OHS—INPT	7	2	1	1	8	1	0	20	1	8
TAVR (O.R.)—INPT	0	0	3	3	0	0	0	6	0	3
Bariatrics—INPT	2	0	0	0	0	0	0	2	0	4
Cardiology—INPT	0	0	0	0	0	0	0	0	0	1
ENT—INPT	0	1	1	0	0	0	0	2	0	2
ENT—OPT	1	2	0	4	0	0	0	7	0	4
AICD and PACERS—INPT	8	2	3	2	2	0	0	17	2	10
AICD and PACERS—OPT	0	5	6	0	5	0	0	16	0	6
Eye—OPT	1	8	4	0	0	0	0	13	0	10
General—INPT	2	5	6	3	4	1	1	22	2	10
General—OPT	0	9	10	10	2	0	0	31	0	10
GU—INPT	4	2	0	0	1	0	0	7	0	5
GU—OPT	0	2	4	4	4	0	0	14	0	4
Neuro/spine—INPT	3	0	2	0	3	0	0	8	0	3
Neuro/spine—OPT	0	1	1	0	0	0	0	2	0	1
Ortho—joints—INPT	8	3	2	2	8	0	0	23	2	8
Ortho—other—INPT	1	1	1	1	2	1	0	7	1	3
Orthopedics—OPT	3	3	4	15	3	0	0	28	3	15
Plastic—INPT	0	0	0	0	1	0	0	1	0	3
Plastic—OPT	1	0	2	0	0	0	0	3	0	3
PMT—OPT	0	0	15	14	3	0	0	32	0	15
Thoracic—INPT	0	0	1	0	4	1	0	6	0	4
Thoracic—OPT	0	0	0	2	0	0	0	2	0	2
Vascular—INPT	12	5	6	3	3	3	1	33	3	12
Vascular—OPT	0	5	5	5	1	0	0	16	0	5
All other—INPT	1	0	1	0	0	0	0	2	0	1
TAVR (CATH lab)—INPT	0	0	2	2	0	0	0	4	0	2
Total no. of cases	54	56	80	71	54	7	2	324		
Total no. INPT. cases	48	21	29	17	36	7	2	160		
Total no. OPT. cases	6	35	51	54	18	0	0	164		

Table 12.4   Beds required by day of week

By category	M	T	W	TH	F	S	SU	
OHS—INPT	27.3	26.0	24.1	27.1	27.2	24.5	26.6	
TAVR (O.R.)—INPT	2.3	2.9	5.6	7.1	5.5	3.7	3.1	
Bariatrics—INPT	2.0	0.0	0.0	0.0	0.0	0.0	0.0	
Cardiology—INPT	0.0	0.0	0.0	0.0	0.0	0.0	0.0	
ENT—INPT	0.5	1.4	1.3	0.5	0.4	0.4	0.3	
AICD and PACERS —INPT	12.6	11.3	10.4	9.7	7.2	5.2	11.0	
General—INPT	13.0	15.6	17.7	17.2	16.8	14.2	12.6	
GU—INPT	6.1	5.7	4.5	4.2	4.0	3.0	3.6	
Neuro/spine—INPT	7.8	7.5	7.7	7.7	8.3	7.0	7.5	
Ortho—joints— INPT	10.9	12.3	11.9	9.0	13.5	10.9	7.9	
Ortho—other— INPT	5.0	5.0	5.3	5.8	6.3	5.7	5.1	
Plastic—INPT	0.7	0.6	0.6	0.9	1.3	1.2	0.9	
Thoracic—INPT	6.4	6.0	5.2	6.6	8.2	7.6	7.1	
Vascular—INPT	29.5	28.7	27.1	25.1	23.4	21.5	25.1	
All other—INPT	1.5	1.6	1.8	1.4	1.0	0.8	1.1	
TAVR (CATH lab) —INPT	1.5	1.9	3.7	4.7	3.7	2.5	2.0	
Total beds req'd.	127.0	126.6	126.8	127.0	126.9	108.3	113.8	856.3
	127	127	127	127	127	109	114	$122.3 \leq 127$
								5.352 = LOS

Table 12.5 Volume

	M	T	W	TH	F	S	SU	Total	Target	LOS	PT days
OHS—INPT	7	2	1	1	8	1	0	20	20	9.14	182.8
TAVR (O.R.)—INPT	0	0	3	3	0	0	0	6	6	5.02	30.1
Bariatrics—INPT	2	0	0	0	0	0	0	2	2	1.00	2.0
Cardiology—INPT	0	0	0	0	0	0	0	0	0	3.60	0.0
ENT—INPT	0	1	1	0	0	0	0	2	2	2.35	4.7
ENT—OPT	1	2	0	4	0	0	0	7	7	0.00	0.0

(continued)

Table 12.5   (continued)

	M	T	W	TH	F	S	SU	Total	Target	LOS	PT days
AICD and PACERS—INPT	8	2	3	2	2	0	0	17	17	3.96	67.4
AICD and PACERS—OPT	0	5	6	0	5	0	0	16	16	0.00	0.0
Eye—OPT	1	8	4	0	0	0	0	13	13	0.00	0.0
General—INPT	2	5	6	3	4	1	1	22	22	4.87	107.1
General—OPT	0	9	10	10	2	0	0	31	31	0.00	0.0
GU—INPT	4	2	0	0	1	0	0	7	7	4.43	31.0
GU—OPT	0	2	4	4	4	0	0	14	14	0.00	0.0
Neuro/spine—INPT	3	0	2	0	3	0	0	8	8	6.69	53.5
Neuro/spine—OPT	0	1	1	0	0	0	0	2	2	0.00	0.0
Ortho—joints—INPT	8	3	2	2	8	0	0	23	23	3.32	76.3
Ortho—other—INPT	1	1	1	1	2	1	0	7	7	5.46	38.2
Orthopedics—OPT	3	3	4	15	3	0	0	28	28	0.00	0.0
Plastic—INPT	0	0	0	0	1	0	0	1	1	6.19	6.2
Plastic—OPT	1	0	2	0	0	0	0	3	3	0.00	0.0
PMT—OPT	0	0	15	14	3	0	0	32	32	0.00	0.0
Thoracic—INPT	0	0	1	0	4	1	0	6	6	7.84	47.1
Thoracic—OPT	0	0	0	2	0	0	0	2	2	0.00	0.0
Vascular—INPT	12	5	6	3	3	3	1	33	33	5.47	180.5
Vascular—OPT	0	5	5	5	1	0	0	16	16	0.00	0.0
All other—INPT	1	0	1	0	0	0	0	2	2	4.63	9.3
TAVR (CATH LAB)—INPT	0	0	2	2	0	0	0	4	4	5.02	20.1
Total cases/week	54	56	80	71	54	7	2	324	324		856.3
										5.352	122.33378

Table 12.6 Daily O.R. minutes used

	M	T	W	TH	F	S	SU	
OHS—INPT	2527	722	361	361	2888	361	0	
TAVR (O.R.)—INPT	0	0	720	720	0	0	0	
Bariatrics—INPT	180	0	0	0	0	0	0	
Cardiology—INPT	0	0	0	0	0	0	0	
ENT—INPT	0	95	95	0	0	0	0	
ENT—OPT	110	220	0	440	0	0	0	
AICD and PACERS— INPT	1200	300	450	300	300	0	0	
AICD and PACERS— OPT	0	625	750	0	625	0	0	
Eye—OPT	55	440	220	0	0	0	0	
General—INPT	470	1175	1410	705	940	235	235	
General—OPT	0	1260	1400	1400	280	0	0	
GU—INPT	780	390	0	0	195	0	0	
GU—OPT	0	220	440	440	440	0	0	
Neuro/spine—INPT	750	0	500	0	750	0	0	
Neuro/spine—OPT	0	160	160	0	0	0	0	
Ortho—joints—INPT	1520	570	380	380	1520	0	0	
Ortho—other—INPT	190	190	190	190	380	190	0	
Orthopedics—OPT	360	360	480	1800	360	0	0	
Plastic—INPT	0	0	0	0	240	0	0	
Plastic—OPT	165	0	330	0	0	0	0	
PMT—OPT	0	0	525	490	105	0	0	
Thoracic—INPT	0	0	165	0	660	165	0	
Thoracic—OPT	0	0	0	240	0	0	0	
Vascular—INPT	1860	775	930	465	465	465	155	
Vascular—OPT	0	625	625	625	125	0	0	
All other—INPT	90	0	90	0	0	0	0	
Total O.R. minutes used	10,257	8127	10,221	8556	10,273	1416	390	≤ 10,277
O.R. minutes available	10,277	10,277	10,277	10,277	10,277	10,277	10,277	

ranges stipulated. Lastly, Table 12.6 indicates the O.R. time that is required by the optimum schedule, by category, each day of the week versus the stipulated time constraints. Again, note that all constraints are satisfied. An analysis of this data yields the fact that 3 operating rooms need not be staffed on Tuesdays, nor 2 on Thursdays. Because no elective surgery is scheduled Saturday or Sunday, the hospital will certainly make less available minutes than what is indicated in the Table. By using the same availability as Monday–Friday in the application, not only allows the application to run as intended, but also provides management with an indication of how the O.R. should be staffed on weekends.

Table 12.7 Percent of patients requiring CTICU beds each day, by day, during LOS assuming all post-Op OHS and TAVR Pts go to CTICU

	+6, +13	Op, +7, +14	+1, +8, +15	+2, +9, +16	+3, +10, +17	+4, +11, +18	+5, +12, +19			
OHS	0.144	1.126	0.905	0.461	0.282	0.196	0.166			
TAVRs		1.000								
		CTICU beds used with schedule								
		M	T	W	TH	F	SA	SU		
		11.249	10.747	12.831	11.245	12.454	10.663	6.411	≤ 13	13
		12	11	13	12	13	11	7		

Lastly Table 12.7 indicates the number of Cardiac Thoracic Intensive Care Beds (CTICU) that will be utilized by day of week.

To summarize and reiterate, the key to engineering the best possible O.R. schedule for any hospital is to utilize the historical relationship that exists between each category of surgery and its length of stay distribution.

The success of deriving the very best O.R. schedule for a hospital rests with its surgeons. If they are willing to adjust existing block schedules, there will be no problem. However, if they are not willing to adjust their existing block schedules, the flow of patients will remain dysfunctional, the hospital will experience grid-lock on days when there are simply not a sufficient number of beds to accommodate all inpatients and will, most assuredly, experience variations in patient days, day-to-day, such that the difference in the number of inpatient days each day and the number of floor beds will equal the number of patients that are accommodated each night on non-floor nursing units; the Emergency Department, the Ambulatory Surgery areas and Post Anesthesia Care Unit.

An Optimum Elective Schedule exists for every hospital. Its derivation lies buried in hospital data. Extract the appropriate data, apply it as indicated and implement the resulting schedule to eliminate inpatients overnight on non-floor nursing units.

Reference

Calichman, M. V. (2005), Creating an Optimal Operating Room Schedule. AORN Journal, 81: 580–588.

Printed in the United States
By Bookmasters